Active LEARNING
Curriculum for Excellence

Third Level
SCIENCE

Ian Cameron

CONTENTS

CONTENTS

HOW TO USE THIS BOOK

Leckie & Leckie's Active Learning series has been developed specifically to provide teachers, students and parents with the ability to implement A Curriculum for Excellence as effectively as possible. Each book has been written with the following objectives in mind:

- To support the implementation of Curriculum for Excellence in schools
- To engage students by taking a contextualised learning approach and providing a range of rich task activities
- To assist teachers with the planning and delivery of lessons and assessment

COVERING CURRICULUM FOR EXCELLENCE

Active Science comprehensively covers the Third Level Outcomes and Experiences for Science, incorporating biology, chemistry and physics.

The Outcomes and Experiences have been organised into chapters, with a different topic covered on each double page spread. The Outcome and Experience reference number for each topic appears in the Table of Contents.

Every double page spread opens with a knowledge summary for a particular topic which conveys key ideas and concepts. The key topic knowledge is enhanced by a practical example, illustration or case study to reinforce learning. A *Top Tip* is also included to highlight key information.

ADDRESSING THE PRIORITIES OF CURRICULUM FOR EXCELLENCE

Active Science focuses on methods to implement the philosophy underpinning Curriculum for Excellence. It addresses the Curriculum for Excellence in a thoroughly practical way that makes learning both engaging and fun!

- Ideas for rich task activities are provided for every topic to enable pupils to gain experiences as well as learning outcomes
- Paired and group learning activities encourage students to take responsibility for and direct their own learning, whilst developing the four capacities
- Creative ideas are offered for making cross-disciplinary links with other classroom subjects, both within the Science Faculty and more widely, to help ensure students join-up their learning

- The relevance of each topic to everyday life is highlighted in order to help students transfer their skills and knowledge to other areas of their lives
- Extension activities are included in the Did You Know? boxes to encourage students' interest and develop breadth and depth of learning

The toolkit of ideas, subject links and activities contained in *Active Science* can be used in the classroom and/or at home. Please see page 7 for details of the text design and key features of the book.

Each double page spread includes the following features:

Activities column. This column contains questions to assess knowledge and understanding as well as rich tasks to deepen understanding of each topic.

Make the Link box. This box highlights the relevance of the topic to a number of other school subjects. This enables learners to gain a more holistic understanding of each topic.

Our Everyday Lives box. This box provides an example of how each topic relates to real life, in order to demonstrate its practical relevance

Did You Know? box. This box contains an additional fact about each topic to engage further interest and to bring the subject to life. It can also be used as an extension activity to broaden and deepen learning.

ASSESSMENT

At the end of each unit, *Active Science* contains an assessment checklist for each topic based around the Third Level Outcomes and Experiences. It also provides:

- Self-assessment mind maps to encourage students to reflect on their learning and their development of the four capacities
- Ideas for inter-disciplinary project work
- Rich task activities around each unit to bring learning to life
- A data handling activity to assist with numeracy and develop students' familiarity with presenting scientific information

SUPPORTING TEACHERS AND STUDENTS

Active Science sets out to provide teachers and students with a valuable toolkit of easy-to-implement ideas for incorporating the philosophies of Curriculum for Excellence into teaching and learning.

The highlighted links with other subjects, activities ideas and real life examples are the perfect starting point for teachers and students to build upon and develop as they explore ideas around a topic. At Leckie & Leckie, our intention is that this *Active Science* book will inspire learners to investigate subjects both widely and deeply in practical and creative ways.

Leckie & Leckie's Active Learning series - bringing Curriculum for Excellence to life.

KEY FEATURES

Each double page spread contains the following features (see below). In addition, at the end of each unit, there are even more activities as well as tools to help track progress.

MAKE THE LINK
draws out links between subjects on a particular topic to aid interdisciplinary learning and deepen understanding

DID YOU KNOW?
boxes provide interesting and engaging facts about each topic to help build knowledge

OUR EVERYDAY LIVES
illustrates how the knowledge translates into practical examples drawn from real life

ACTIVITES TO TRY
Short and engaging revision questions and rich tasks

KEY WORDS
Most important words and phrases highlighted in bold

EXAMPLE
Examples given are captivating and spark students' interest

FULL-COLOUR
Bright and stimulating colour throughout

TOP TIP
Key facts and concepts are highlighted to aid knowledge retention

ANIMALS AND THEIR HABITATS

WHAT IS A HABITAT?

A habitat is a place or an area where animals live. In its habitat an animal must be able to find food, water, shelter, a mate and anything else it needs to survive. There are many different types of habitat, and animals have evolved over thousands of generations to survive in a huge variety of places. Camels and polar bears have each adapted in special ways so that they can live and breed in their very different habitats. Not all habitats are on land – coral reefs are an example of a marine habitat.

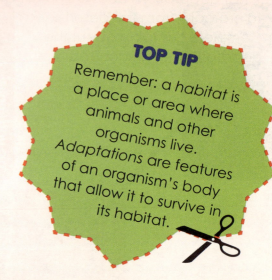

TOP TIP

Remember: a habitat is a place or area where animals and other organisms live. Adaptations are features of an organism's body that allow it to survive in its habitat.

ADAPTATIONS OF CAMELS

Camels can survive for long periods of time without food and drink. Their humps are a reserve of energy-rich fat and their bodies use water very efficiently, producing little sweat and very concentrated urine. When they are thirsty they can drink over a hundred litres of water at one time. To protect them from sand they have very long eyelashes and ears lined with hair, and they are able to close their nostrils. Since they eat mostly plants they have no need for claws or pointed teeth. Camels have thick rubbery lips so that they can eat dry prickly plants, and large feet so that they do not sink into soft sand.

ADAPTATIONS OF POLAR BEARS

Polar bears live and hunt on and around the Arctic ice. To help them survive in such a cold area their bodies are insulated by a layer of blubber which can be 10 cm thick. Under the white fur they have black skin which absorbs as much energy as possible from the weak rays of the Arctic sun. Their coat consists of two layers and the outer 'guard' hairs are hollow and transparent. Small ears and a small tail help to reduce heat loss. Their white fur provides perfect camouflage in the ice and snow and makes it easy for them to sneak

up on their prey. They have a very good sense of smell and can detect a seal over 1 km away or buried under nearly 1 m of snow. Large feet spread their weight on soft snow and their soles have small bumps to help grip the ice. Polar bear claws can be up to 5 cm long and are more curved than on other bears. This helps them to catch and hold prey, grip the ice and dig. Large paws also help make them excellent swimmers – they can swim at speeds of nearly 10 km per hour and have been seen as far as 100 km from land.

ACTIVITIES

QUESTIONS

1. What adaptations does a camel have to protect it from the desert sand?
2. Complete this table about the adaptations of polar bears.

Adaptation	Purpose
Layer of blubber	
	Absorb heat from sun
	Grip on ice
Curved claws	
Large feet	

3. Raptors like eagles can spot their prey from a great height. They then swoop to kill and eat the prey. What adaptations do you think they might have?
4. Predators which hunt other animals usually have eyes on the front of their heads. The prey which are likely to be caught and eaten usually have eyes on the sides of their head. Why do you think they have adapted in this way?

AND NOW TRY

1. Search for distribution maps for polar bears and camels.
2. Find out which animal species are endangered in Scotland. Design a poster or a web page to raise awareness of the problem.
3. Imagine you have bred a Scottish wildcat in captivity and want to reintroduce it to the wild. What are some of the problems you might have? Discuss in a group and devise an action plan.
4. Plants can become endangered too. Why are they becoming endangered? Are the reasons the same as for endangered animal species? Write an article for a school newspaper looking at some of these issues.

MAKE THE LINK

Modern Studies – you learn about politics and conservation. **Geography** – you learn about different regions and habitats. **PSE** – you learn about environmental issues and how you can contribute to conservation efforts.

DID YOU KNOW?

It's estimated that over 5 000 species of animal and 25 000 species of plant are currently facing extinction.

OUR AMAZING WORLD

9

The giant panda is probably the best known example of an endangered species. Habitat loss is one of the main reasons they have become endangered. Their only food is bamboo which grows at altitudes between 500 and 3 100 metres, and a breeding pair of pandas needs 30 square kilometres of land to sustain them. Their forest habitat has been lost due to people using land for agriculture, bamboo harvesting, herb collection and industrial development. China Conservation and Research Center for the Giant Panda at Wolong http://www.chinagiantpanda.com/site/wolong.htm is the most important panda-breeding facility in the world. In 1986 the first panda was born in captivity. Now there are over ten cubs born each year and a programme is being developed to reintroduce captive-bred animals to the wild. Many zoos across the world feel they have a role to play in conservation through captive breeding of endangered species.

PHOTOSYNTHESIS

MAKING FOOD

Plants capture light energy with their leaves. They then use this energy together with water from the soil and carbon dioxide from the air to make food. The process is called **photosynthesis** and is summarised in the picture below.

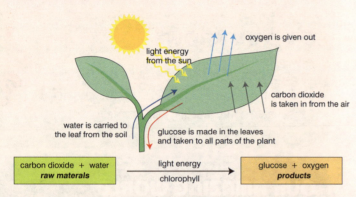

carbon dioxide + water *raw materals*	light energy chlorophyll →	glucose + oxygen *products*

Photosynthesis makes glucose and gives out oxygen. Another way of expressing this is in a word-equation like this:

$$\text{water + carbon dioxide} \xrightarrow{\text{light}} \text{glucose + oxygen}$$

PHOTOSYNTHESIS AND ANIMALS

Unlike plants, animals cannot make their own food so they must obtain their energy by eating plants or other organisms. All of the energy originally comes from the sun. Light energy is captured in the process of photosynthesis and stored in the plants as chemical energy. When animals eat the plants the energy is then passed to them.

The energy-transfer process looks like this:

Light energy from the sun → converted to chemical energy in plants → eaten by animals to provide energy.

More about this can be found in the section on food chains on page 100.

RESPIRATION

Inside the cells of their bodies all living things release the chemical energy stored in their food by a process called **respiration.** In the process of respiration chemicals from food combine with oxygen to release energy. The word-equation is similar to that for burning fuels:

glucose + oxygen → carbon dioxide + water + energy

BALANCE

Plants make their food by photosynthesis. They take in **carbon dioxide** and give out **oxygen**.

Animals get their energy from their food by **respiration**. They breathe in **oxygen** and breathe out **carbon dioxide**.

The two process balance each other. Or do they? More about this in the section on climate change.

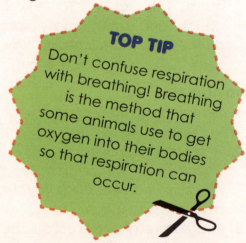

TOP TIP

Don't confuse respiration with breathing! Breathing is the method that some animals use to get oxygen into their bodies so that respiration can occur.

10

ACTIVITIES

QUESTIONS

1. What is the energy change in photosynthesis?
2. Photosynthesis requires light. What are the other two things which are needed?
3. What are the two products of photosynthesis?
4. How do plants get their food?
5. How do animals get their energy?
6. What is the name of the process that all living things use to get energy?

AND NOW TRY

1. Set up your own terrarium (see **Our amazing world** for more information). All you need is a large bottle, you will find lots of instructions online at www.leckieandleckie.co.uk.
2. Try growing sprouting seeds such as mung beans in a jar. What happens if you keep them in the dark?
3. Have look at NASA's website (http://science.nasa.gov/) and search for information about plants in space.

MAKE THE LINK

Geography – you'll learn about the value of plants in converting carbon dioxide to oxygen. Carbon dioxide is a greenhouse gas, meaning that it contributes to climate change (see pages 28). So the more trees and plants there are, the better for the world climate and the world environment.

Home Economics – you learn about types of food. A measure of how much energy is stored in food. The more calories, the more energy there is stored in the food.

Biology – photosynthesis is a key topic in biology courses.

DID YOU KNOW?

The sun provides the energy for all processes on the planet, and indeed the whole solar system. To provide this energy the sun's mass decreases by 4 million tonnes each second. And it will last for another billion years.

11

OUR AMAZING WORLD

NASA astronauts are experimenting with growing plants in space. This might make it possible to fly much longer missions. The crew need food and oxygen and, at present, they have to take supplies of both to last for the whole mission. A supply of live plants would provide food and also oxygen by photosynthesis, and at the same time use up carbon dioxide breathed out by the people.

CHEMICALS AND AGRICULTURE

FOOD FOR PLANTS

In the previous section we looked at photosynthesis and saw how plants make their food using water, carbon dioxide and the energy from the sun. As often happens in science when we look more closely, the situation is actually a bit more complicated. It is possible to get plants to grow with only carbon dioxide and water but only for a short time, as they really need other minerals and compounds for healthy growth. The main requirements are nitrogen, phosphorus and potassium. The plants obtain these from the soil through their root systems. In this section we will concentrate on nitrogen.

PROTEINS

All living things need nitrogen. They need it to make proteins. Proteins are the building blocks of life. All living things need proteins to build and repair cells and to store energy.

NITROGEN

Although the earth's atmosphere is nearly 80% nitrogen, plants cannot use this directly for growth. They need the nitrogen in the soil to be **fixed** into **compounds** which they can absorb. The roots of certain plants, like clover and beans, contain nitrogen-fixing bacteria in small nodules on their roots. Bacteria in these nodules combine atmospheric nitrogen with oxygen and hydrogen to form compounds like ammonium and nitrates which can be used by plants.

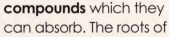
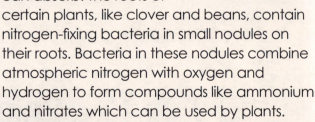

Root

Nodule

THE NITROGEN CYCLE

Nitrogen compounds in the soil are taken up by plants to make proteins. These proteins are then eaten by animals. Animal waste and the remains of dead plants and animals are then broken down by bacteria and fungi and the nitrogen compounds return to the soil. Once in the soil they can again be absorbed by plants. This process is called the nitrogen cycle.

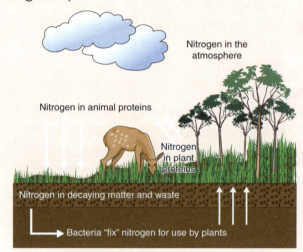

Nitrogen in the atmosphere

Nitrogen in animal proteins

Nitrogen in plant proteins

Nitrogen in decaying matter and waste

Bacteria "fix" nitrogen for use by plants

FERTILISERS

In order to produce more food and feed the growing population, people have used fertilisers to increase the nitrogen in the soil and so increase crop growth. This can be done by the use of animal waste such as manure. It can also be done by manufactured chemical fertilisers. There are some problems:

• animal manure and urine release methane, nitrous oxides, ammonia and carbon dioxide. All of these are greenhouse gases responsible for climate change. Some chemical fertilisers also release greenhouse gases.

• the nitrogen compounds in chemical fertilisers are not all taken up by the plants. Some are washed away by rainfall. When they reach the rivers, lakes and oceans they reduce the level of dissolved oxygen in the water and so reduce its ability to sustain life.

• chemical fertilisers also affect the soil's structure and stability.

12

ACTIVITIES

QUESTIONS

1. Name three elements essential for plant growth.
2. Why can atmospheric nitrogen not be used by plants directly?
3. Give two ways in which nitrogen compounds are fixed in the soil.
4. How is the nitrogen in animal protein returned to the soil?
5. What is the main problem with the intensive use of animal fertilisers?
6. What are the main problems associated with chemical fertilisers?

AND NOW TRY

1. The crop rotation diagram below refers to catch crops. Find out what is meant by the phrase 'catch crop'.
2. The problems of feeding the world's population are immense. The Sustainable Food Laboratory is an international organisation whose mission is to promote sustainable food sources for the planet. Have a look at their website. http://www.sustainablefoodlab.org/. Design a poster and write a letter to an MP to raise awareness of this issue.
3. Seeds like cress can be grown on moist paper with no nutrients. Try growing some, then compare the effect of growing them on compost.
4. People with allotments and gardens also rotate their crops. Research a crop rotation plan for a garden or an allotment.

TOP TIP

All plants need nitrogen for healthy growth. Nitrogen compounds can be added to the soil by using animal waste or chemical fertilisers.

MAKE THE LINK

English – you may read *The Grapes of Wrath* by John Steinbeck, which tells the story of people leaving their homes in the 1930s as a result of the dust bowl. Think about how the dust bowl relates to how the land is used.

History – you learn about the depression during the 1930s.

Geography – you learn about sustainable land use and crops.

Music – the Bakersfield sound is a genre of country music which the people from the plains states took to California when they were forced to leave their homes.

DID YOU KNOW?

During thunderstorms, lightning flashes can also produce compounds of nitrogen which are useable by plants.

13

OUR AMAZING WORLD

Crop rotation is the practice of growing a variety of crops in sequence on the same piece of land. Here is a sample crop rotation plan:

This is done for a number of reasons:

- to prevent the build-up of harmful bacteria and pests in the soil
- to avoid using up all the soil's nutrients by one type of crop
- to prevent soil erosion

You will notice that the cycle includes the planting of clover to increase the nitrogen content of the soil.

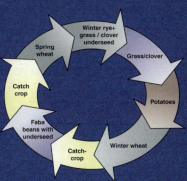

ENERGY TRANSFER AND BUILDINGS

CONDUCTION, CONVECTION AND RADIATION

Heat energy will always flow from a hot object to a cooler object. This can happen in three ways.

CONDUCTION

In conduction, the heat energy travels through a solid material. The particles of the solid cannot move (though they can vibrate). The energy is transferred from one to another and so passes through the material.

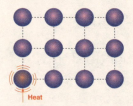

Heat

In general metals are good conductors of heat and non metals are good insulators.

CONVECTION

Heat travels upwards through liquids and gases by convection.

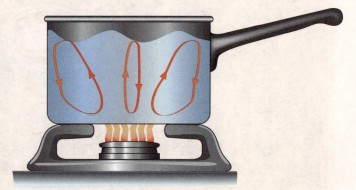

The particles in liquids and gases are able to move past each other so the hot particles at the bottom rise and the cooler ones at the top fall. This creates convection currents.

RADIATION

Radiation transfers the heat energy in the form of waves. This means that the energy can travel through empty space. The heat energy from the sun reaches us by radiation.

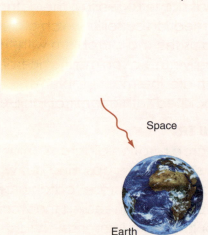

Space

Earth

Dark surfaces absorb and emit radiation much more effectively than light or shiny surfaces. Light or shiny surfaces tend to reflect rather than absorb the energy. You will notice that on a sunny day you feel much more comfortable in light clothing than in dark. This is because the dark colours absorb the sun's heat energy more effectively.

TOP TIP

Conduction takes place in solids. Metals are the best conductors of heat. Convection carries heat upwards in liquids and gases. The liquid or gas circulates to form a convection current. Radiation transfers heat in the form of waves. Dark surfaces are the best emitters and absorbers of radiation.

ACTIVITIES

QUESTIONS

1. What are the three methods by which heat energy is transferred?
2. What sort of surfaces absorb and emit most radiation?
3. Why is it more comfortable to wear white on a sunny day?
4. What name is given to the natural convection currents which glider pilots use?
5. Give three ways of reducing heat loss from a house.
6. How does heat travel from the sun to earth?

AND NOW TRY

1. On a sunny day you may see gulls and other birds soaring in thermals, which are areas of warm air rising upwards away from the surface of the earth in a convection current. The birds use thermals to gain height in the same way a glider pilot does. You can tell because, just like a glider, the birds go round in circles, getting higher and higher without flapping their wings. You can watch them while you are outside on a sunny day.
2. Carry out a free home energy check at http://www.energysavingtrust.org.uk/proxy/view/full/165/homeenergycheck
3. Fill a drink bottle with cold water, wrap it in a black bin liner and leave it in the sun for an hour or two. How hot does it get?
4. Your home or a relative's home may qualify for a government grant to increase its insulation. Have a look at http://www.freeinsulation.co.uk/
5. Find out what is meant by the phrase 'carbon foot print'.

MAKE THE LINK

Geography – you learn about winds. Lots of these are caused by convection. You also learn about the Gulf Stream. This is in fact a convection current which occurs on a global scale.

DID YOU KNOW?

In Scotland we usually want our homes to be warmer than the air outside. So heat energy will naturally flow from the warmer house to the cooler surroundings.

- Heat travels through the walls, floor, roof and windows by conduction.
- In the loft, convection currents carry the heat up to the roof.
- Heat can be lost from walls, roof and windows by radiation.

Reducing heat losses from our homes means that we use less energy. This also saves money and reduces our 'carbon foot print'.

15

OUR AMAZING WORLD

On warm days convection currents occur naturally over dark-coloured fields, buildings and sunny hillsides. These are called thermals and glider pilots use them to gain height. The glider is initially launched by a tug plane or a winch. The pilot circles in the first thermal to gain height, then loses height as he glides to the next thermal where he circles again to climb. By repeating this process it is possible to travel hundreds of miles and climb to heights of over 10 000 metres.

RENEWABLE ENERGY

ENERGY FROM THE SUN

All the energy on earth came originally from the sun. Even the non-renewable energy stored in fossil fuels came originally from the sun. After all, fossil fuels were once plants and microbes.

The sun's energy not only provides **heat**, it also drives the earth's **wind** and all our weather systems and, along with the gravitational effect of the moon, causes the movement of **tides** in the oceans.

When we talk about renewable energy, we are talking about capturing these forms of energy. They do not produce any pollution or greenhouse gases and they will last for ever. But there are complications and people have valid concerns about the environmental impact of these energy sources. At present, it is more expensive to harness these forms of renewable energy than to burn fossil fuels, so countries and companies are reluctant to invest. In Scotland we have set a target to produce 50% of our energy from renewable sources by 2020.

HYDROELECTRIC POWER

To generate hydroelectic power, water (from rain and rivers) is captured in a dam above the power stations. As it falls to a lower level its **potential energy** is converted into electricity:

Potential energy → kinetic energy → electrical energy.

This can have an impact on the environment. Large areas of land are flooded when dams are built and huge tunnels are often dug through mountains to move the water.

WIND TURBINES

Wind turbines capture the kinetic energy of the wind and convert it to electrical energy. They are a major part of Scotland's drive to increase the use of renewable energy and are increasingly being built offshore in areas like the Solway Firth.

Not everyone approves. The turbines are often over 100 m high and some argue that these and the pylons to carry the electricity spoil the beauty of our countryside. Others are concerned about their impact on wildlife.

TIDAL ENERGY

In some areas around Scotland the tides can flow at rates of up to four metres per second and the water level rises and falls by up to 6 m. Scientists are working to develop better ways of harnessing the energy of the tides and also of the waves. Tidal turbines are in place in the seas around Orkney and Shetland.

People are concerned about these in the same ways as they are about wind turbines.

SOLAR WATER HEATING

matt black solar panels on roof

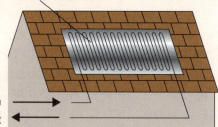

cold water in
hot water out

Panels on the roofs of houses can capture the sun's energy and use it to heat water. Notice how they are dark-coloured to absorb as much radiation as possible. You will see these panels on the roofs of houses somewhere near you and they are commonly used to heat water for swimming pools in holiday resorts.

PHOTOVOLTAIC CELLS

Photovoltaic cells capture the sun's energy and convert it to electricity. The William Rankine building at Edinburgh University has been built with its entire south-facing wall covered in these cells. They are expensive to produce and often used in remote areas where there is no other electricity supply.

ACTIVITIES

QUESTIONS

1. What is the source of all the earth's energy?
2. What energy changes take place in a hydroelectric power station?
3. What energy change takes place in a photovoltaic cell?
4. What name is given to the capture of heat by buildings?
5. What causes the tidal movements in the oceans?
6. Give two reasons why people are concerned about wind turbines.

AND NOW TRY

1. Make a poster to explain what is meant by the idea of embodied energy.
2. Find out how long a wind turbine and a photovoltaic panel take to repay their embodied energy (see **Our everyday lives**).
3. Write a letter to an MP giving your views on offshore wind turbines.
4. Do some research on tides and find the location of the strongest tidal streams around Scotland.

TOP TIP
Renewable energy sources rely on the sun's energy that is continuously reaching the earth. They will last forever. Non-renewable sources will run out because the energy they store is finite – it will not last forever.

MAKE THE LINK

Business Education – you learn about supply and demand and how long it takes for an investment to make a profit.

Modern Studies – you learn about the developing economies of the Pacific Rim and their need for energy.

Geography – you learn about the oceans and the landscape.

DID YOU KNOW?

The damming of rivers has an effect on wildlife, and when this was done after World War Two people were worried about the impact. Writing about this period author Emma Wood quotes someone's worries:

'I heard about drowned farms and hamlets, the ruination of the salmon-fishing and how Inverness might be washed away if the dams failed inland.' (Emma Wood, *The Hydro Boys: Pioneers of Renewable Energy*. Luath Press, Edinburgh, 2004)

In China the Three Gorges hydroelectric scheme flooded over 600 square kilometres of land.

17

OUR AMAZING WORLD

Embodied energy

Energy is needed to manufacture everything, whether it's a car or a toothbrush. Energy is needed to extract the materials needed, to make the article and then to recycle it at the end of its useful life. This energy is called the **embodied energy**.

It is important that the things we build, like wind turbines provide us with more energy than it takes to make them. They must repay their embodied energy.

FOSSIL FUELS

The term 'fossil fuel' refers to coal, oil and natural gas. We call them fossil fuels because they are made from the remains of plants, animals and micro-organisms that lived over a hundred million years ago. They are non-renewable energy sources, which means they will eventually run out. One of the challenges facing us is to reduce our use of these fuels and conserve them for future generations.

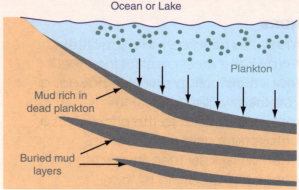

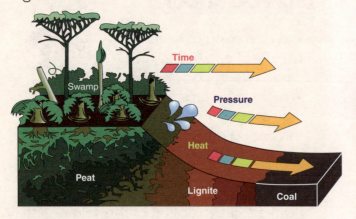

COAL

Coal was formed from the remains of large tree-like plants which died and were covered by layers of mud and earth. Over millions of years the material became buried more deeply and the action of heat and pressure formed first peat, then lignite (a type of brown coal) and finally coal.

OIL AND GAS

Oil and gas were formed from the remains of tiny plants and animals called plankton which live in the oceans. As with coal, when they died they were buried by mud and over millions of years oil and gas were formed.

Because oil and gas would naturally tend to come up to the surface, they are only found in special rock formations where they are trapped in the spaces of porous rock, a bit like water in a sponge, and underneath solid rock.

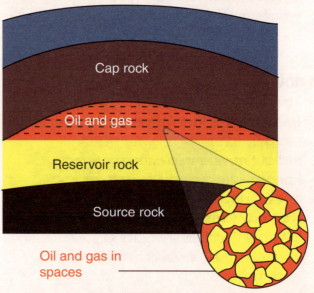

The major oil companies of the world spend vast amounts of time and money looking for these rock formations.

HOW LONG WILL THEY LAST?

New reserves of fossil fuels are still being found, so it is very difficult to estimate how much longer they will last. At the present rate of consumption it is likely that we will have used up the earth's supply of oil and gas in less than 50 years, while coal supplies may last for another 150 years. Currently, over 80% of the world's energy comes from fossil fuels. Supplies will begin to run out and as they do the price of these fuels will rise. This will lead to more investment in renewable energy sources. See the previous section.

18

ACTIVITIES

QUESTIONS

1. What are the three main fossil fuels?
2. Which is the most plentiful fossil fuel?
3. Why is it difficult to estimate how long fossil fuels will last?
4. What will happen to the price of fossil fuels as they start to run out?
5. When were fossil fuels formed?
6. What is the meaning of 'non-renewable'?
7. What prevents the oil and gas rising to the earth's surface?

AND NOW TRY

1. Scientists are now developing 'clean coal' technology. Find out what this term means and where it has been applied to build a power station.
2. Make a poster explaining how acid rain is produced.
3. Can China develop its economy without damage to the environment? What should the developed countries of the West be doing to help? Discuss in groups.
4. Find out about diseases caused by people breathing smoke from burning coal and those that affected coal miners.

MAKE THE LINK

History – the industrial revolution, which began in Britain in the eighteenth century, pioneered the use of fossil fuels to power machinery.

Business Education – you learn about the pressures of supply, demand and price.

Modern Studies – you learn about the developing economies of the world and how they are adding to the world's energy use.

DID YOU KNOW?

Apart from the need to conserve the limited supply of fossil fuels there are some environmental issues.

Burning fuel uses oxygen and produces energy and combustion products like water, carbon dioxide and sulphur dioxide. We can summarise it like this:

Fuel + Oxygen → heat energy + carbon dioxide + water vapour + sulphur dioxide

As we will see in the section on climate change, carbon dioxide is one of the main gases responsible for the greenhouse effect (see page 28 for the section on climate change). To reduce the effects of climate change we must produce less carbon dioxide, which means we must burn less fossil fuel.

19

OUR AMAZING WORLD

Coal is the most plentiful of the fossil fuels and in the UK we have supplies to last some hundreds of years. Burning coal is one of the main causes of acid rain because it produces lots of sulphur dioxide and fine soot particles (particulates). During the industrial revolution of the nineteenth century, vast quantities of coal were mined and it was used to power factories, heat homes and drive ships.

TOP TIP

The three fossil fuels are coal, oil and natural gas. They were formed millions of years ago from dead plants and animals and cannot be replaced.

CHANGES OF STATE

THE STATES OF MATTER

We saw in the last section how the particles are arranged in the three states of matter: solids, liquids and gases. When water evaporates to become steam we have a change of state: liquid to gas. When it freezes to become ice the change of state is liquid to solid. The most common changes of state are summarised in the diagram below.

SOLID → (Melting) → LIQUID → (Boiling) → GAS
GAS → (Condensing) → LIQUID → (Freezing) → SOLID

SUBLIMATION AND DEPOSITION

While the four changes of state above explain most of the things we see, there are two other possibilities. Some substances change directly from solid to gas when they are heated, and this process is called **sublimation.** Solid carbon dioxide is often called dry ice and turns directly back into a gas, with quite dramatic effects.

The opposite of sublimation is **deposition,** where a gas turns directly to solid. This happens on frosty nights when the water vapour in the air turns directly to ice, giving us the beautiful hoarfrost effect.

ENERGY AND TEMPERATURE AND THE EARTH

Heat energy supplied to a material causes its temperature to rise. Heat energy from the sun warms up the planet. The energy does not affect all parts in the same way.

To raise the temperature of water requires about five times as much energy as is needed to produce the same temperature rise in soil or rock. This means that in summer the land heats up much more quickly than the oceans and in the winter it cools down much more quickly. So in a small area of land, like Scotland, surrounded by water, the temperature ranges from about –5°C to 25°C.

In the centre of a large land mass like the plains of Canada, which are thousands of miles from the sea, the temperature range is much greater. Winter temperatures are often below –30°C and summer days are often much hotter than in the UK. The oceans of the earth play a vital role in keeping the temperature stable.

The processes can be summarised like this:

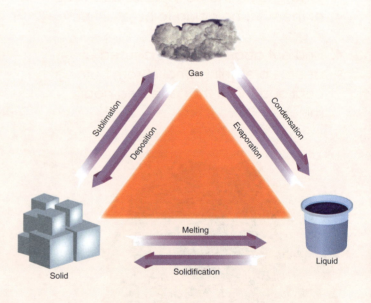

20

ACTIVITIES

QUESTIONS

1. What name is given to the following changes of state:
   a) Solid to liquid?
   b) Liquid to solid?
   c) Liquid to gas?
   d) Gas to liquid?
   e) Solid to gas?
   f) Gas to solid?
2. Why do small land masses heat up more slowly in the summer?
3. What is formed when water vapour in the air turns directly to ice in winter?
4. Why is the temperature range in Scotland less than in Canada?

AND NOW TRY

1. What are the long-term consequences of the shrinking of the polar ice caps? Do some research and design a poster.
2. Find out how changes of state are used in cold packs and refrigerators. Have a look at home.howstuffworks.com
3. In the 1960s a high school student in Tanzania discovered that when he put hot water in the freezer it turned to ice before cold water. This is called the Mpemba effect. Find out about the Mpemba effect and try to carry out an experiment at home to see if he was right.

MAKE THE LINK

Geography – you learn about climate in different parts of the earth.

Home Economics – you learn about preserving food by freezing.

Art & Design – natural effects like hoarfrost can inspire your photography or drawing.

DID YOU KNOW?

If you leave a kettle boiling with the lid off, the water will take a very long time to evaporate. This is because it takes a very large quantity of energy to change water into steam.

When you put an ice cube in your drink it does not melt immediately. To melt it must first absorb a large amount of energy from the drink. This is why your drink cools down so much.

Changes of state need energy – lots of it.

21

OUR AMAZING WORLD

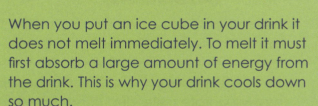

In the summer the ice in the north of the planet starts to melt. It freezes again in winter. These changes need very large quantities of energy. So the polar ice caps are vital in keeping the temperature of the planet stable.

There is evidence that the ice caps are getting smaller each year. This not only causes sea levels to rise, it also makes the earth's climate less stable.

HEATING AND COOLING MATTER

ENERGY AND PARTICLES

There are three states of matter: solid, liquid and gas. All are made up of tiny particles called **atoms** and **molecules**. These particles are constantly moving or vibrating, and when the matter is heated the heat energy makes the particles move faster.

SOLIDS

In a solid the particles are closely bound together. Imagine a set of marbles held together with springs.

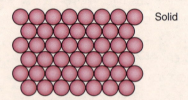
Solid

The marbles, like the particles in a solid, cannot move from their fixed position but they can vibrate around their fixed position. If heat provides particles in the solid with more energy they will vibrate faster and need more space. The solid will **expand.** If the solid is cooled down the particles will move more slowly and take up less space. The solid will **contract.**

LIQUIDS

In a liquid the particles are free to move past each other. The spacing is about the same as in a solid.

Just as with a solid, heating will cause the particles to move faster and the substance will **expand.** Cooling will make the particles move more slowly and the substance will **contract.**

Liquid

GASES

In a gas the particles are much further apart and moving rapidly about at random. Imagine a large number of small children running around the playground.
The particles of the gas move in the same way.

Gas

The explanation is exactly the same as for solids and liquids. Heating causes the particles to move faster and further apart and the gas **expands.** Cooling causes them to move more slowly and the gas **contracts.**

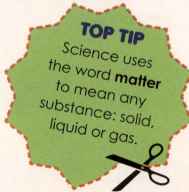
TOP TIP
Science uses the word **matter** to mean any substance: solid, liquid or gas.

22

ACTIVITIES

QUESTIONS

1. Name the three states of matter.
2. What is the effect of heat on the particles in matter?
3. What happens to all matter when it is heated?
4. What happens to all matter when it is cooled?
5. When you put water in a freezer to make ice does the ice take up:
   a) More space than the water?
   b) Less space than the water?
   c) About the same space as the water?
6. Think about our answer to question 5. Does this mean the particles in the water are:
   a) Further apart than in the ice?
   b) Closer together than in the ice?
   c) About the same distance apart as in the ice?
7. When a kettle boils and fills the room with steam, what happens to the space between the particles of water in the steam? Give a reason for your answer.

AND NOW TRY

1. If you have a jar where the lid is tight, like a jam jar, try running it under the hot tap. Why do you think this helps?
2. Sometimes it goes wrong and a bridge collapses due to expansion or contraction of the metal used to build it. Try an internet search for bridges that have collapsed due to expansion and contraction.
3. Try taking an empty plastic drink bottle, a big one, and filling it with hot water from the tap. Empty out the hot water – this will leave the air inside nice and warm. Now put the top on tight and run it under the cold tap. Can you explain what happens?
4. Try an internet search for 'anomalous expansion of water'.

MAKE THE LINK

Geography – expanding and contracting air causes sea breezes; expanding and contracting sea water causes ocean currents. **CDT** – buildings and structures have to be designed to allow for expansion and contraction.

DID YOU KNOW?

A large structure like the Forth Rail Bridge will expand in summer and contract in winter. Its total length can change by up to one metre. Clearly the engineers who design bridges have to allow for this and it can be done in a number of ways. You may see expansion

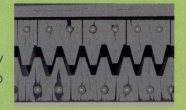

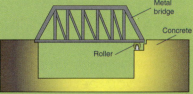

joints like this in the roadway of a bridge. They allow the bridge to expand and contract without damage. Or the ends of the bridge may rest on rollers.

23

OUR AMAZING WORLD

Water is one of the commonest substances on the planet but it has some very unusual properties.

- When water is cooled below 4°C it starts to expand instead of contracting. This is why ice forms on the top of water when it freezes, not the bottom.
- When water freezes into ice the volume increases slightly. This explains why, when water freezes in pipes, it sometimes bursts them.

AIR PRESSURE

TONNES OF AIR

We live at the bottom of a huge sea of air. There is the equivalent of 10 tonnes of air pressing down on each square metre of the earth's surface. This is **air pressure** or **atmospheric pressure.**

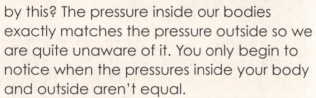

300km

So why aren't we crushed by this? The pressure inside our bodies exactly matches the pressure outside so we are quite unaware of it. You only begin to notice when the pressures inside your body and outside aren't equal.

When you are in an aeroplane taking off, the pressure outside your body reduces slightly. Your ears may hurt or pop because the pressure inside you body is now greater than that outside. Sucking a sweet, chewing gum or swallowing helps balance the pressures and you feel OK again.

Sometimes the pressure of the air has a very dramatic effect. The people operating this tanker made a mistake in the sequence of closing valves. It was left with no air inside and the pressure of the air outside crushed it.

Suction pads are held in place by air pressure. When you stick them to a surface you push out the air between the sucker and the glass. The pressure of the air outside then holds them in place.

WHAT CAUSES AIR PRESSURE?

We looked at how particles are arranged in solids, liquids and gases in the previous section. The particles in a gas are well spaced and constantly moving.

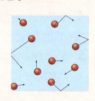

The particles are moving fast – several hundred metres per second – so when they strike a surface there is a force on that surface. The force of these collisions causes air pressure. Scientists call this explanation the **kinetic theory of gases.**

ACTIVITIES

QUESTIONS

1. Why does the pressure of the air not crush our bodies?
2. When an aeroplane lands, your ears may hurt. How does the pressure outside your body now compare with the pressure inside?
3. What do you call a space with all the air removed?
4. Spray cans like deodorant warn that they should not be exposed to heat. What do you think would happen to the pressure inside if they were heated?
5. Can you think of any other situations where your body may feel increased pressure?

AND NOW TRY

1. Look at the weather forecast and the synoptic chart in a newspaper or online. When the pressure rises what weather can we expect? What weather can we expect when the pressure falls?
2. If you have a barometer in your house or in school, record the barometric pressure every day for a few weeks. Can you become your own weather forecaster?
3. Try this trick: place an old wooden ruler or a thin piece of wood so it sticks halfway over the edge of a bench. Smooth a newspaper on top of it and squeeze out all the air. You should now be able to break the ruler by hitting it with a hammer or a mallet. The pressure of the air holds the paper in place.
4. Try a video search for 'railroad tank implosion'

MAKE THE LINK

Geography – you learn about weather systems and how they are affected by atmospheric pressure; you also learn about contour lines on maps, which are similar to the isobars on a synoptic chart.

CDT – you learn about the strength of structures.

DID YOU KNOW?

As we move higher – in a plane or up a mountain – the air pressure decreases. This is because there is less air higher up to press down on you. An aircraft altimeter uses this fact to measure the plane's height.

OUR AMAZING WORLD

25

Atmospheric pressure does not remain exactly the same all the time. The slight changes in pressure can be measured with a barometer and are important in weather forecasting. Meteorologists – scientists who study and forecast weather – produce maps showing the pressure at different points on the earth and how it is likely to change over the next few days. These maps are called synoptic charts and the lines on them which join points of equal pressure are called isobars. These charts are a vital tool in weather forecasting.

TOP TIP

Remember, the particles in a gas are constantly moving. Air pressure is caused by the fast-moving air particles striking a surface.

GASES OF THE AIR

WHAT IS AIR?

Air is a mixture of gases. It is mostly made up of oxygen and nitrogen, with a very small amount of argon and carbon dioxide.

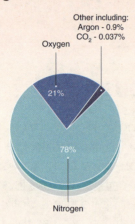

Other including:
Argon - 0.9%
CO_2 - 0.037%

Oxygen

21%

78%

Nitrogen

OXYGEN

Oxygen is the gas which allows things to burn. When most fuels burn to give us energy, carbon combines with oxygen to form carbon dioxide.

carbon (in fuel) + oxygen → carbon dioxide + water + energy

Oxygen gas is colourless, with no smell or taste. It can be identified by the **glowing splint test.**

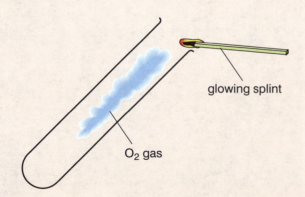

glowing splint

O_2 gas

A burning wooden splint is blown out but left glowing. When placed in the test tube of oxygen it relights.

CARBON DIOXIDE

Carbon dioxide is breathed out by most organisms and is also produced by burning fuels. The balance between the production of carbon dioxide and its use by plants is explained in the sections on **photosynthesis** and **climate change.**

Carbon dioxide does not allow things to burn and it is identified by the **lime water test.**

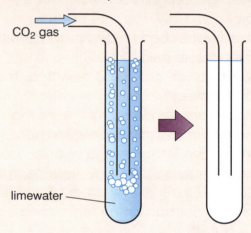

CO_2 gas

limewater

Lime water is a clear liquid, it is added to the test tube and shaken. If carbon dioxide is present the lime water turns cloudy because tiny particles of chalk are formed.

NITROGEN

Nitrogen is just there. It makes up most (almost four fifths) of the air that we breathe. It can be identified from what it doesn't do. It doesn't allow anything to burn and it doesn't affect lime water. In fact, it takes part in very few chemical reactions. We make use of the fact that nitrogen is unreactive. It can be used to surround the filaments in light bulbs to stop them burning. Nitrogen is the gas that fills bags of crisps to keep them fresh, and it is now sometimes used to inflate car tyres to increase their life and efficiency.

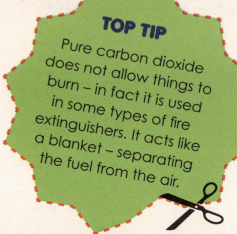

TOP TIP

Pure carbon dioxide does not allow things to burn – in fact it is used in some types of fire extinguishers. It acts like a blanket – separating the fuel from the air.

26

ACTIVITIES

QUESTIONS

1. Which gas is used up when things burn?
2. Which gas is produced when fuels burn?
3. Which test is used to identify oxygen?
4. Which test is used to identify carbon dioxide?
5. Why is nitrogen sometimes used to inflate car tyres?
6. What is the meaning of the term 'unreactive'?
7. Is oxygen a reactive or an unreactive gas? Give a reason for your answer.

AND NOW TRY

1. Try lighting a candle in a shallow dish of water and then covering it with a glass.
   Can you explain what happens?
2. At high altitudes in the mountains the air is sometimes said to be 'thinner'. What do you think this means? Some athletes have tried to make use of this in their training. Find out what is involved in 'altitude training'.
3. Have a look at a fire extinguisher if you have one at home. A number of chemicals are used and some of them contain carbon dioxide.
4. Have a look at the contents list on a can of fizzy drink. Which gas is in the bubbles?

TOP TIP

Make sure you remember the two gas tests. Oxygen relights a glowing splint. Carbon dioxide turns lime water cloudy.

MAKE THE LINK

PE – you learn about oxygen uptake during exercise.

Geography – you may learn about the oil industry and divers working in the North Sea.

Home Economics – you learn about energy from food and how to use this knowledge to keep you fit and healthy.

DID YOU KNOW?

Argon makes up nearly 1% of the air and is a member of a family of elements called the **noble gases.** The other members of the family are helium, neon, krypton, xenon and radon. They are all highly unreactive so they take part in almost no chemical reactions. Argon is also used to surround light bulb filaments.

27

OUR AMAZING WORLD

Helium is well known as the gas used to make lighter-than-air balloons. It is also sometimes used in diving. People who go diving for pleasure usually breathe from a tank of compressed air, but when commercial divers are working at great depths and high pressures the normal gases in the air can become poisonous. In cases like this the divers sometimes use heliox – a mixture of helium and oxygen – and it makes them speak in high squeaky voices.

CLIMATE CHANGE

THE GREENHOUSE EFFECT

If you have been in a greenhouse or have a conservatory, you will have noticed that it can get very hot on a sunny day. This is because the energy from the sun passes in through the glass. If you were outside most of the energy would be reflected away again but in a greenhouse a lot of it is trapped by the glass, as shown in the picture. This makes the temperature much higher than it is outside.

SUN'S RAYS

GLOBAL WARMING

Remember, plants take in carbon dioxide and give out oxygen. Animals take in oxygen and breathe out carbon dioxide. The two processes of photosynthesis and respiration are in balance. In recent years mankind has been burning a lot of fossil fuel, which releases carbon dioxide into the atmosphere. So over the last few decades the amount of carbon dioxide in the atmosphere has increased. The logic is simple:

More carbon dioxide → greenhouse effect of atmosphere increases → more of the sun's energy is trapped → temperature of the planet increases

Almost all scientists agree that this is happening and that the dangers are:
• rising sea levels (as polar ice caps melt)
• large areas of land becoming desert
• many species becoming extinct

New evidence is being collected all the time so no one is sure how big the problem really is.

The same thing can happen on a global scale.

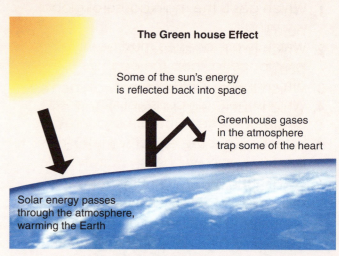

The Green house Effect

Some of the sun's energy is reflected back into space

Greenhouse gases in the atmosphere trap some of the heart

Solar energy passes through the atmosphere, warming the Earth

This is exactly like the greenhouse or the conservatory, only this time the energy is not trapped by the glass but by gases in the earth's atmosphere – mainly carbon dioxide. This has been happening for much of the life of the planet. It is not new, and for centuries the planet has stayed roughly the same temperature.

TOP TIP
The greenhouse effect is the trapping of the sun's energy by gases (especially carbon dioxide) in the earth's atmosphere. Increased carbon dioxide is the main cause of global warming. Other causes are: methane, water vapour and oxides of nitrogen.

ACTIVITIES

QUESTIONS

1. Which gas is the main cause of global warming?
2. Which two processes must be in balance to keep the earth's atmosphere stable?
3. What is the main cause of increased carbon dioxide in the atmosphere?
4. Is the greenhouse effect something that is caused by human activity? Give a reason for your answer.
5. Large areas of forest have been cleared to make way for industry and farming. How will this affect the level of carbon dioxide in the atmosphere?

AND NOW TRY

1. Methane gas is another important cause of climate change. Find out what it is and how it is produced.
2. There is a government-sponsored website, http://campaigns.direct.gov.uk/actonco2, which allows you to calculate how much carbon dioxide you and your family produce. Visit and have a try.
3. Plan a trip to London travelling from your home town. You could fly or go by train. Compare:
   a) the cost
   b) the journey time, from the time you leave your house till the time you get to the hotel in London
   c) the carbon dioxide produced
4. Find out what is meant by carbon offsetting.
5. Look at the website of the Intergovernmental Panel on Climate Change http://www.ipcc.ch/.

MAKE THE LINK

Geography – you learn about climate and the factors affecting it.

Modern Studies – you learn how governments are addressing climate change.

PSE – you learn about responsible citizenship and environment issues.

Religious, Moral and Philosophical Education – you may learn about different beliefs about the planet and our responsibilities to it.

DID YOU KNOW?

Some scientists disagree with the global and UK predictions for climate change. They are sometimes called 'climate sceptics'. They argue that the planet's climate has changed before, and that predicting climate change and its effects is very difficult. They believe that the causes of climate change cannot easily be isolated to one factor such as an increase in carbon dioxide emissions.

29

OUR AMAZING WORLD

Carbon capture and storage is a suggested way of reducing greenhouse gases. The idea is that the carbon dioxide from burning fuels would be captured and stored underground in old oil fields and coal mines. This would only be done for large producers of carbon dioxide, like power stations.

Not everyone agrees that this is a good idea. There is always the worry that the gas would escape and bubble out into the sea. If this were to happen, not only would we have the carbon dioxide back but it might slightly change the pH of the water and create a whole new set of problems.

THE SOLAR SYSTEM

NINE PLANETS

The solar system is made up of nine planets, all travelling in orbits round the sun. The sun is a star very similar to lots of other stars in the sky, but much closer. Stars are huge balls of hot gas and generate their heat by nuclear reactions. There are two types of planets in our solar system. The terrestrial or rocky planets are Mercury, Venus, Earth, Mars and Pluto. The gas planets are, as the name suggests, huge balls of gas. These are Jupiter, Saturn, Neptune and Uranus, and are often called gas giants.

HOW BIG IS IT?

'Space is big. Really big. You just won't believe how vastly hugely mind-bogglingly big it is.' (Douglas Adams, *The Hitchhiker's Guide to the Galaxy*. Pan Macmillan, London, 1979.)

It is almost impossible to imagine the distances in the solar system, and the pictures tend to give the impression that the planets are much closer to each other than they really are. Let's build a model.

• We will use a football to represent the sun.
• The diameter of the sun is about a hundred times the diameter of the earth, so we will use a pea for the earth.
• To correctly represent the distance between the sun and the earth the pea needs to be 25 metres away from the football.

Suppose we want to add Pluto to our model.

• Pluto is much smaller than the earth so we will use a pinhead to represent it.
• And to represent the distance to Pluto the pinhead would have to be about 1 000 m away from the football.

MOONS

As the earth orbits the sun the moon orbits the earth. So it looks a bit like the picture below – but remember, the pictures always make them look closer than they really are.

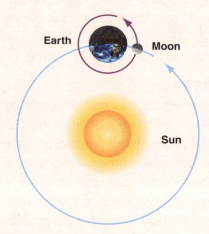

Many of the other planets also have moons; in fact, Jupiter has more than 50 moons. Most of them are very small, but four – Europa, Ganymede, Io and Callisto – are large and were first observed in 1610. In relation to its size, earth has the largest moon.

TOP TIP
Remember the mnemonic. **M**y **V**ery **E**asy **M**ethod **J**ust **S**peeds **U**p **N**aming **P**lanets gives us the first letters of the planets in order: Mercury, Venus, Earth, Mars, Jupiter, Saturn, Uranus, Neptune and Pluto.

ACTIVITIES

QUESTIONS

1. What are the two different types of planet?
2. What name is given to the paths of the planets round the sun?
3. Which planet was demoted to a dwarf?
4. The earth orbits the sun. Which object orbits the earth?
5. What is the problem with most of the pictures of the solar system?
6. How does the sun generate its heat?

AND NOW TRY

1. Try looking at http://www.northern-stars.com/solar_system_distance_scal.htm for an idea of the scale of the solar system.
2. There is far too much information about the solar system to fit in these pages. For more knowledge a good place to start looking is NASA's website: http://solarsystem.nasa.gov/planets/index.cfm
3. Have a look at the night sky and find which planets you can see. There are lots of websites that tell you what to look for at different times of year. Try Jodrell Bank. http://www.jodrellbank.manchester.ac.uk/astronomy/nightsky/
4. Watch the phases of the moon for a month and make sketches as it goes through its phases. Can you find an explanation for what is happening?

MAKE THE LINK

Geography – you learn about the earth and its place in the solar system.

RME – you learn some of the stories and myths that gave the planets their names.

English – you may read some science fiction set on other planets.

DID YOU KNOW?

In 2003 another object in the solar system was discovered and thought to be a tenth planet. It was given the name Eris. After a lot of discussion the International Astronomical Union decided it was too small to be a planet and is really a dwarf planet – along with a lot of other objects in the solar system. They also agreed that Pluto is really a dwarf planet, leaving the solar system with only eight planets!

31

OUR AMAZING UNIVERSE

Is there life elsewhere in the solar system? Earth is a rocky planet, so the first place to look would be the rocky planets. Mercury is hot enough to melt lead and Pluto is too cold to support life. Venus is covered in clouds of sulphuric acid with a surface temperature of 500°C, so that leaves Mars. It has always been the favourite among science fiction writers for intelligent life. Most scientists now think that if there were any life there it would be very primitive. Mars Global Surveyor found evidence of water, and it has been suggested that some simple form of life like bacteria might have existed there. Some people have suggested there might be some primitive life on Titan, which is one of Saturn's moons.

BEYOND THE SOLAR SYSTEM

UNDERSTANDING THE SIZE

Let's go back to our model where the sun is represented by a football and the Earth by a pea. The pea was 25 m away from the football.

- We will have another football to represent the nearest star.
- To correctly represent the distance to the nearest star, the second football has to be 6 000 km away. It has to be further away than New York.

LIGHT YEARS

We need a new system of measuring to deal with the huge distances involved. We measure in a unit called light years. Light years measure distance. We think of distance in terms of how long the journey would take.

Here are some examples of how far things can travel in one second.

Moving object	Distance travelled in 1 second
Sprinter	10 metres
A car on the motorway	35 metres
A jet airliner	250 metres
Sound	340 metres
Light	300,000,000 metres

In one second light can travel seven times round the Earth.

A light year is the distance light can travel in one year.

When we think of distances in terms of the time the light takes to reach us we get numbers like this:

Object	Time for light to reach us
Sun	8 minutes
Nearest star	4 years
Andromeda (nearest galaxy)	200 million years

So we say the nearest star is four light years away and Andromeda is 200 million light years away.

GALAXIES

A galaxy is a group of stars: about a million million stars. Our earth orbits the sun. The sun is a star in a galaxy called the Milky Way. The Milky Way is about 100 000 light years across. If we could leave our galaxy, the next one we would come to is called Andromeda, and it is 200 million light years away. All the galaxies make up the universe.

top view of our galaxy

100 000 light years

32

ACTIVITIES

QUESTIONS

1. Arrange these in increasing order of size: galaxy, planet, universe, solar system.
2. How long does light take to reach us from the sun?
3. How long does light take to reach us from the nearest star?
4. Which is the nearest galaxy to us?
5. How far is it across our galaxy?
6. What is a light year?

AND NOW TRY

1. Have a look at http://www.universetoday.com/2009/12/18/the-known-universe-video/
2. As with the section on the solar system, there is far too much information to fit this page. NASA's website is a good place to start searching for more details: http://solarsystem.nasa.gov/planets/profile.cfm?Object=Beyond
3. We see different star systems at different times of year. Again, the Jodrell Bank website is a good place to start looking for information.
4. Stars in the sky are not all the same colour. The hottest ones like Rigel are bluish and cooler ones like Betelgeuse are reddish. If you look to the south in winter you can see these two stars the constellation of Orion.

TOP TIP

Remember: a light year is a distance. It is the distance light can travel in one year. In fact, it is nearly 10 000 000 000 000 km, but that doesn't really help. The only way to deal with it is to think in terms of the speed of light.

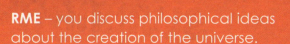

MAKE THE LINK

RME – you discuss philosophical ideas about the creation of the universe.

English – you may read the legends that gave constellations their names, or science fiction about space travel.

DID YOU KNOW?

There are many types of stars in the galaxy, all at different stages of their lives. Some are very dense and some just a cloud of gas called a **nebula**. The densest known object is a black hole. The gravitational field of a black hole is so strong that nothing can escape, not even light. At the centre of our Milky Way galaxy is a black hole called Sagittarius A* (pronounced 'A-star') which is four million times heavier than the sun.

33

OUR AMAZING UNIVERSE

Most scientists describe the origin of the universe as the 'Big Bang'. About thirteen billion years ago all the material that makes up the universe started out as one huge hot mass that exploded and expanded outwards. The matter in the universe is still expanding outwards today. All the galaxies in the universe are moving away from each other. Observing this constant expansion of the universe is what led to the Big Bang theory in 1929. If all the galaxies are moving away from each other, then in the past they must have been closer together ... so a long time in the past they were all in the same place ... so there must have been a big bang to separate them.

SUMMARY ACTIVITIES

WORLD POPULATION

WORLD POPULATION

This project on population includes links to maths, history and geography. You will look at how developments in science affected the earth's population, and link this to events in history. Geography is related because it is about where people live and how they use the land, and you will need maths skills to collect data and draw a graph.

In the middle of the fourteenth century the world's population was about 450 million. The Black Death, or bubonic plague, killed about half of Europe's population around 1350 (see the section on micro-organisms). Since then the population of our planet has grown steadily. Look at the table for some approximate figures. They have been taken from various sources and have been rounded. You may want to carry out your own research to check these figures.

Year	World population (millions)
1500	400
1550	400
1600	400
1650	450
1700	500
1750	750
1800	1000
1850	1250
1900	1600
1950	2600
2000	6000

Using a long sheet of paper such as a roll of wallpaper, make a big graph with these figures. It will end up looking something like this:

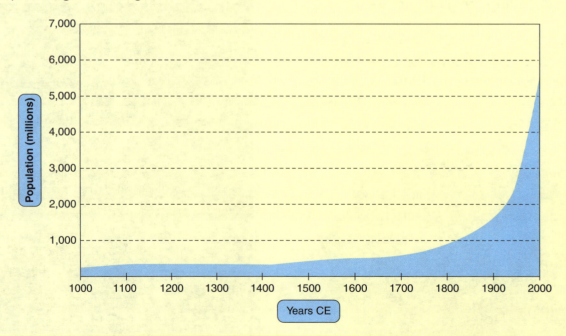

34

ACTIVITIES

1. Match the following key words with their meanings:

KEYWORDS:
a) photosynthesis
b) habitat
c) nitrogen fixing
d) fertiliser
e) conduction
f) convection
g) renewable energy
h) condensation
i) star
j) galaxy

MEANINGS:
1. an area where plants or animals live
2. the transfer of heat in solids
3. a ball of hot gas
4. the forming of nitrogen compounds in the soil
5. energy sources which will not run out. e.g. solar, wind, tidal energies
6. a group of millions of stars
7. the change of state from a gas to a liquid
8. the capture of the sun's energy by green plants
9. a chemical added to soil to increase its nitrogen content
10. the transfer of heat upwards in liquids or gases

2. The table shows the energy sources used to generate electricity in the UK. Display this data in a pie chart like the one shown below. Remember to colour in the sectors and to add a key.

Source	Percentage
Coal	35
Gas	37
Nuclear	21
Renewable	7

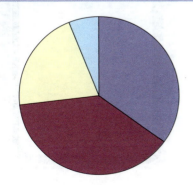

3. The following are some events that seriously affected the world population. Mark them on a timeline, and add some pictures and some text about them:

- From about 1750 the agricultural revolution brought about changes in food production, and the industrial revolution meant that people moved from the country to towns.
- Scientific advances: the introduction of antiseptics by Joseph Lister and the discovery of penicillin by Alexander Fleming reduced death rates and helped the population to grow.
- World War One and the flu epidemic that followed it caused millions of deaths.

Find at least four other scientific and medical advances that may have affected the population and add them in. Now add at least six historical events. For example: the discovery of America; the Act of Union between Scotland and England; the French Revolution; and other things you may have learned in history.

Complete the following mindmap.

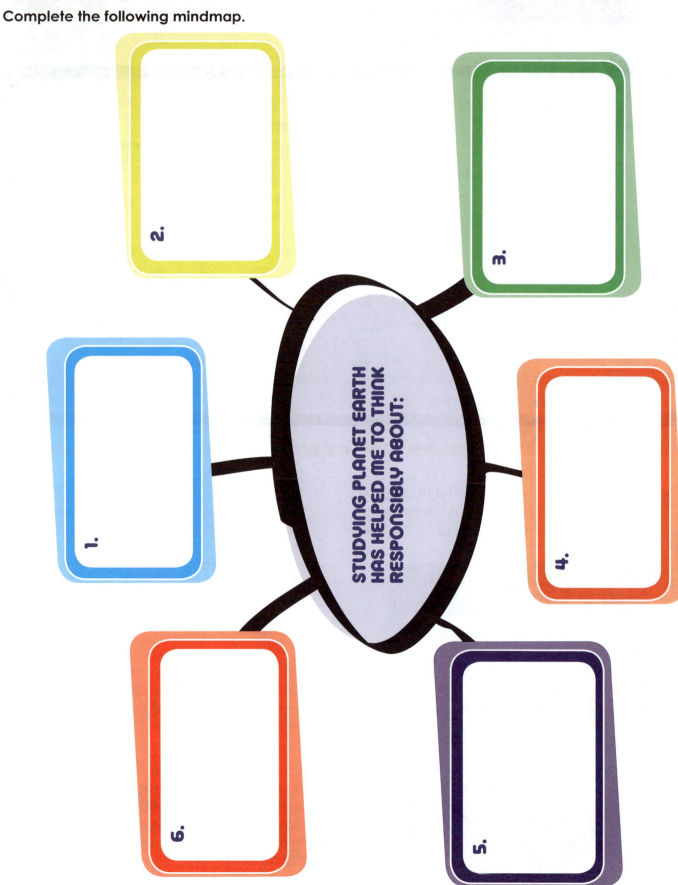

STUDYING PLANET EARTH HAS HELPED ME TO THINK RESPONSIBLY ABOUT:

1.

2.

3.

4.

5.

6.

36

MY PROGRESS

MY LEARNING CHECKLIST

	Help needed	Getting there	Good to go!
1. I can identify living things from different environments and explain how they have adapted for survival.	⬭	⬭	⬭
2. I have investigated the process of photosynthesis and understand its importance to life on earth.	⬭	⬭	⬭
3. I can explain the use of chemicals in agriculture and their importance to world food production.	⬭	⬭	⬭
4. I can use my knowledge of heat transfer to contribute to the design of energy efficient buildings.	⬭	⬭	⬭
5. I can discuss the benefits and problems of different renewable energy sources.	⬭	⬭	⬭
6. I can explain changes of state in terms of particles and explain their importance in nature.	⬭	⬭	⬭
7. I can explain the process of climate change and its effect on living things.	⬭	⬭	⬭
8. I have developed an understanding of the structure of the universe and can discuss the prospect of space travel and the possibility of life existing elsewhere.	⬭	⬭	⬭

37

FRICTION

ROUGH SURFACES

Friction is a force which opposes all motion. It always acts in the opposite direction to the motion and tends to slow down or stop a moving object. In the diagram below the block is moving to the left so the frictional force acts to the right.

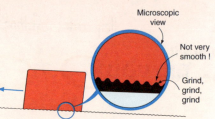

Microscopic view

Not very smooth!

Grind, grind, grind

The surfaces of the block and the table top may look smooth, but if we examined them with a microscope we would find that they are rough, with lots of little irregularities. As they move, these rough surfaces catch on each other and cause the frictional force. Whenever there is movement friction always causes some of the kinetic energy to change into heat. We use this effect when we rub our hands together to warm them up.

If the object were on wheels the frictional forces would be greatly reduced but there would still be friction between the wheels and the axles and between the wheels and the road.

REDUCING FRICTION

Friction will be present in anything with moving parts, whether it's a train or an electric toothbrush. Friction will oppose the motion and waste some of the energy by turning it into heat. It will also wear moving parts as they rub over one another. Engineers and designers always try to reduce friction in machinery. The commonest way to do this is by lubrication. As the chain passes over the cogs some of the energy the cyclist provides is wasted and turned into heat. The chain is oiled to lubricate it, this reduces the frictional force and less energy is wasted.

A thin layer of oil between the surfaces keeps them apart and reduces friction.

Designing suitable lubricants for different situations is an important part of technology.

AIR CUSHION VEHICLES

Keeping the moving surfaces apart with a layer of oil reduces the friction. Sometimes they are kept apart with a cushion of air. The vehicle floats or 'hovers' above the track because of the air that is being blown out of the small holes. This greatly reduces friction.

Hovercraft and air hockey games make use of the same idea.

HELPFUL FRICTION

Friction opposes all motion, but without friction life would be impossible. Every time you pick something up you depend on the frictional forces between the object and your hand. When you walk, the frictional force between your shoes and the floor means your leg muscles can push you forward. When you want to stop, you need that frictional force again. Think how difficult it is to walk on an icy pavement or pick things up with soapy hands – the ice and the soap reduce the frictional force and things can get a bit dangerous.

Sometimes things are designed to increase the frictional forces, like the soles of these training shoes.

TOP TIP

Friction occurs whenever two surfaces move over one another. A frictional force always acts in the opposite direction to the motion and always turns some of the kinetic energy into heat.

38

ELECTRICITY AND WAVES

ACTIVITIES

QUESTIONS

1. In which direction does a frictional force act?
2. What energy change is always produced by friction?
3. What causes friction?
4. How can friction be reduced?
5. Give an example of a situation where we would want to reduce friction.
6. Give an example of where friction is useful.
7. Why do cars find it difficult to stop on icy roads?

AND NOW TRY

1. Carry out an experiment with two magnets to see how the fields could be used to make a hover train.
2. Hovercraft are built for use by the armed forces.
   a) Where is the biggest military hovercraft?
   b) How would it be used?
3. Plan a trip on the bullet train http://www.japan-guide.com/e/e2018.html.
4. Try a web search for high friction materials.
   You can find which surfaces have the greatest friction by a simple experiment. Place the object on a flat surface (a large book will do). Now increase the angle till it just starts to slide. The bigger the angle, the bigger the frictional force.

MAKE THE LINK

CDT – you learn about the moving systems and machinery.

Business Education – you learn about emerging technology in the Pacific Rim.

PE – you may need special trainers like the ones pictured for certain sports to help prevent you from slipping.

DID YOU KNOW?

The Japanese Shinkansen or bullet train can travel at 400 km per hour.
It uses the idea of hovering to reduce friction. Engineers are developing technology which will allow the train to hover above the rails using magnetic fields and reach even faster speeds of over 500 km per hour.

OUR AMAZING WORLD

Hovercraft are used all over the world both for carrying passengers and for military purposes. To reduce the frictional forces they float on a cushion of air. Sometimes they are called air cushion vehicles. The diagram below shows how the air is blown downwards and lifts the hovercraft slightly out of the water. The big propellers on top move it forwards and it can travel much faster than a conventional ship.

Propeller

Air

Fan

Flexible skirt

FRICTION AND CARS

We saw in the last section that friction can be both a friend and an enemy. There are times when we want to reduce friction and times when we want to increase it. This is very clear in the design of a motor car. Friction between the tyres and the road means the car can be steered and controlled. If this friction is lost due to ice on the road, then driving becomes very dangerous indeed. Friction in the brakes means the car can stop. On the other hand, some frictional forces simply slow down the car and waste energy. By understanding and reducing these frictional forces, cars can be made more efficient so that they use less fuel.

STREAMLINING

Whenever a car moves there will be air resistance. The air flowing over the car produces a frictional force which opposes its motion. This friction wastes energy, so reducing it will reduce the energy loss and make the car more efficient. Early car designers didn't really bother about this, but modern cars are carefully designed and tested in wind tunnels to reduce their air resistance. Modern cars are said to be more **streamlined**, or more **aerodynamic**.

AIR RESISTANCE AND SPEED

As a vehicle travels faster, the air resistance increases. Doubling the speed causes four times the air resistance. So clearly driving

more slowly reduces the air resistance and saves energy. It has been suggested that if the speed limit on motorways were reduced from 70 mph to 60 mph the average family would make considerable savings in fuel costs.

BRAKES

A car or a bike's braking system depends on the frictional force between brake pads and a disc.

The pads press against the discs to increase the frictional forces, and this causes the car to slow down. Special materials have been developed to make the frictional forces as large as possible and so slow the car down quickly.

TYRES

The frictional force between the tyres and the road makes the car stop when the brakes are applied. It also means the car can accelerate and be steered. If the roads are slippery due to rain, ice or snow the frictional force is very much smaller and the car is more difficult to control. Manufacturers are continually developing new material and new tread patterns to improve the performance of their tyres.

ELECTRICITY AND WAVES

ACTIVITIES

QUESTIONS

1. What is air resistance?
2. In what two places in a car are high frictional forces needed?
3. What can reduce the frictional force between the tyres and the road?
4. What happens to the air resistance force as the speed of the car increases?
5. During a fuel crisis in the 1970s the US government reduced the speed limit on all roads to 55mph. Why do you think they did this?

AND NOW TRY

1. Have a look at the Shell eco-marathon website http://www.shell.com/home/content/eco-marathon-en/welcome_global.html and find the world record for a car's fuel consumption.
2. See http://www.archive.org/details/facts_on_friction for some film about friction in old vehicles.
3. Do some research on car fuel consumption in family cars.
   a) Which has the highest fuel consumption?
   b) Which has the lowest fuel consumption?
4. Should speed limits on our motorways be reduced to save fuel? Make a presentation or organise a debate.
5. Write a letter to your MP suggesting that speed limits be reduced to save fuel. Give your reasons.

TOP TIP
In the design of a car, large frictional forces are needed in the braking system and between the tyres and the road. The frictional forces of the air flowing over the car and the tyres rolling on the road are kept to a minimum.

FORCES, ELECTRICITY AND WAVES

MAKE THE LINK

CDT – you learn about new materials that can be used in construction.

Business Studies – you may hear about initiatives by some big companies to contribute towards more environmentally friendly technologies.

PSE – you learn about road safety.

DID YOU KNOW?

Manufacturers are trying to produce high frictional forces for braking and steering but low frictional forces when the cars is being driven at a steady speed. Reducing what is called the rolling friction when the car is driving along will make the car more efficient and so use less fuel.

41

OUR AMAZING WORLD

When cars first appeared on our roads people were not aware of the need to conserve our fossil fuels. Over the years cars have become much more economical and it is now quite common for family cars to travel 50 miles on a gallon of fuel. This has been achieved in a number of ways. Engines burn the fuel more efficiently, cars are lighter and the frictional forces have been reduced. For the Shell eco-marathon cars were designed which can travel several thousand miles using only one gallon of fuel.

FIELD FORCES

There are hundreds of forces which affect our everyday lives. Most of these are **contact forces**. The force you exert to open the fridge door, to pick up your school bag, kick a ball and dozens of other activities require contact between the two objects. These are called contact forces. When you use a magnet to pick up a pile of pins the magnet does not have to touch the pins. The force field round the magnet pulls the pins towards it. The **magnetic force** is an example of a field force.

MAGNETIC FORCES

Magnets have many uses in our lives, such as closing cupboard doors and holding screws to the ends of screwdrivers. In an electric motor it is the magnetic force that converts the electrical energy into kinetic energy.

There are two other field forces: the **electrostatic force** and the **gravitational force**.

ELECTROSTATIC FORCE

If you rub a plastic ruler or a comb it will attract and pick up small pieces of paper. Again, it does not have to touch them – the electrostatic field pulls them towards it. Photocopiers and inkjet printers use the electrostatic force to make the ink stick to the right place on the paper and so build up an image. At home electrostatic force is what makes cling film stick to itself and to the food you are using it to wrap.

GRAVITATIONAL FORCES

Gravity is the third field force, and it is very much smaller than the other two. We only notice the gravitational force when it involves a very large object, like a planet or a star. The gravitational force between the earth and the sun keeps the earth in its orbit around the sun. In the same way, the gravitational force

between the earth and the moon keeps the moon in its orbit around the earth. There is gravitational attraction between all objects, but it is usually too small to notice. If we place two 1 kg masses about 30 cm apart, the gravitational force between them will be less than a billionth of a Newton, but the gravitational force between the earth and a 1 kg mass is approximately 10 Newtons.

WEIGHT

Everything on or near the earth has a force pulling it towards the earth's centre. We call this force the **weight**. Weight depends on the strength of the gravitational field, and on earth the gravitational field strength is approximately 10 Newtons per kilogram (kg). So the weight pulling down on a 1 kg mass is 10 Newtons (N), the weight pulling on a 2 kg mass is 20 N and on a 7 kg mass is 70 N

weight = mass × gravitational field strength

But gravitational field strength is not the same everywhere. If you travelled to the moon, where the gravitational field is weaker, weight would decrease. On Jupiter, with a much stronger gravitational field, weight would increase. Even the earth's gravitational field strength is not the same in all places. It is slightly greater at the poles than at the equator.

MASS

The mass of an object tells us how much matter it contains, or how many atoms make it up. This does not change. A 1 kg mass will contain the same quantity of matter and the same number of atoms everywhere in the universe. It will always be a 1 kg mass. Its weight, which depends on the strength of the gravitational field, may change. The weight would be more on Jupiter, less on the moon and zero in an orbiting space craft. But the mass will always be 1 kg.

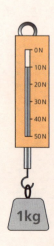

ELECTRICITY AND WAVES

ACTIVITIES

QUESTIONS

1. What are the two groups of forces?
2. Name three field forces.
3. Give an application of the electrostatic force.
4. Give an application of the magnetic force.
5. How does the strength of the gravitational force compare with the electrostatic and magnetic forces?
6. What is the weight of a 5 kg mass on earth?
7. What would happen to your weight if you were to travel to the moon?
8. What name is given to the experience of astronauts when they seem to float in space?

AND NOW TRY

1. Look on You Tube for videos of astronauts in zero gravity.
2. Zero gravity can be simulated by an aircraft flying a curved path. This was the method used to train astronauts. It is possible to book a zero g experience – find out how much it would cost.
3. Find the gravitational field strength for other planets in the solar system.
4. Try rubbing a balloon on your hair and sticking it to the wall, demonstrating the electrostatic force.

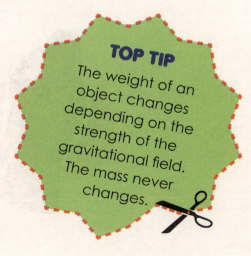

TOP TIP
The weight of an object changes depending on the strength of the gravitational field. The mass never changes.

MAKE THE LINK

Geography – you learn about the structure of the earth.

Business Studies – you may read case studies about oil companies and the search for oil.

Physics – electromagnetic forces are a key topic within physics courses.

DID YOU KNOW?

In an orbiting spacecraft like the shuttle the astronauts and all the other objects seem to be weightless. Objects don't fall, they seem to float. Their mass does not change, but they are weightless because they do not feel the effect of the earth's gravitational field. It always looks like they are having fun, but the crew need extensive training so that they can live and work in these conditions.

43

OUR AMAZING WORLD

In the section on fossil fuels we learned that scientists and technicians are constantly looking for new sources of oil and gas. To do this they must find the special rock structure where these fuels might be trapped. By making very accurate measurements, it is possible to locate places where the earth's gravitational field is slightly stronger than usual. This gives the first indication that there might be oil-bearing rocks in the area. These and other tests make it possible to find the best place to drill test wells.

DENSITY AND FLOATING

The density tells us the mass of one cubic centimetre of a material. For example, one cubic centimetre of water has a mass of 1 gram (g). Density is measured in grams per cubic centimetre (g/cm³), so we say that the density of water is 1 g/cm³.

Water is often taken as a standard and the density of other materials are compared with the density of water. Steel, for example, is nearly eight times as dense as water. One cubic centimetre of steel has a mass of about 7.8 g. Its density is 7.8 g/cm³. Here are some more examples:

Material	Density in g/cm³
wood (spruce)	about 0.5
titanium	4.5
aluminium	2.7
lead	11.3
gold	19.3
air	0.0017

DENSITY AND APPLICATIONS

Clearly, the density of a material determines the use that can be made of it. Fishing weights are often made of a dense metal like lead, while aircraft are built using aluminium, which has a much lower density.

CALCULATING DENSITY

To calculate the density of a material you divide the mass in grams by the volume in cubic centimetres.

$$density = \frac{mass}{volume}$$

FLOATING

Wood usually floats in water. This is because the density of wood (about 0.5 g/cm³) is less than the density of water (1 g/cm³). Aluminium has a density of 2.7 g/cm³ and will sink if placed in water. Things float if they are less dense than the liquid in which they are placed. One liquid can float on top of another. In some salad dressings the less dense oil floats on top of the denser vinegar.

TOP TIP

The density of a material tells us the mass of 1 cm³ of that material. It is calculated from the equation:

$$density = \frac{mass}{volume}$$

For an object to float in water it must be less dense than water.

Example 1
Suppose 200 cm³ of cola has a mass of 210 g.

$$density = \frac{210}{200} = 1.05 g/cm^3$$

Example 2
A piece of wood measures 4 cm × 10 cm × 6 cm and has a mass of 120 g.
To calculate its density we first have to calculate the volume:

Volume = 4 × 10 × 6 = 240 cm³

Then we calculate the density:

$$density = \frac{120}{240} = 0.5 g/cm^3$$

44

ELECTRICITY AND WAVES

ACTIVITIES

QUESTIONS

1. How do you calculate density from mass and weight?
2. The unit for mass is g and for volume is cm³. What is the unit for density?
3. What is the density of water?
4. A certain type of plastic has density 1.2 g/cm³. Will it float in water? Give a reason for your answer.
5. 50 cm³ of alcohol has a mass of 40 g. Calculate its density.
6. A block of wood measures 8 cm × 3 cm × 10 cm. If it has a mass of 180 g:
   a) what is its volume?
   b) what is its density?

AND NOW TRY

1. Lead is a very dense metal, but there are many metals that are more dense. Find out about 'depleted uranium': its density, its uses and its dangers.
2. Steel is more dense than water but steel ships float. Can you explain or find out why?
3. Hot air is less dense than cold air. This is why hot air balloons float upwards in the air. Buy a kit to build your own hot air balloon http://www.otherlandtoys.co.uk/hotairballoon.htm.
4. Eggs normally sink when you put them in water. If you add salt to the water, do you make it more dense so the egg will float?
5. Hydrogen is the least dense gas known. It was used to make airships in the 1930s, but in 1937 one of them caught fire. Search for the Hindenburg disaster on You Tube.

MAKE THE LINK

CDT – you learn about materials and how to match their properties to their uses.

Maths – you learn how to calculate area and volume.

History – you learn about aircraft of the Second World War.

DID YOU KNOW?

One of the stages in making whisky or beer is to ferment a mixture of barley water and yeast in a mash tun. The yeast turns the sugars in the barley to alcohol. Alcohol is less dense than water so when alcohol is produced the liquid becomes less dense. This means the brewer can check the progress of the fermentation by measuring the density.

OUR AMAZING WORLD

The density of the human body is almost exactly the same as that of water. Not too surprising, as our bodies comprise around 65% water. When you are in a swimming pool you may just be able to float lying on your back. It is very much easier to float in the sea. This is because the salt in the sea water makes it more dense than your body. The water in the Dead Sea and in the Great Salt Lake in the US is so salty that people float very high in the water.

BATTERIES

The first battery was produced by Alessandro Volta in 1800 and consisted of a series of silver and zinc plates separated by layers of blotting paper. The blotting paper was soaked in salt water.

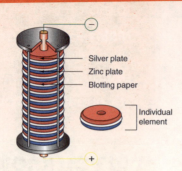

Silver plate
Zinc plate
Blotting paper

Individual element

Today small, lightweight batteries are increasingly important for mobile phones, computers, MP3 players and all the devices we now take for granted. Large batteries which can store lots of energy are needed as we develop electrical cars, and special batteries are being developed to store the electricity generated by wind turbines. The battery industry is an important part of the global economy.

All batteries have one thing in common. They store chemical energy, which is converted to electrical energy when they are used.

chemical energy → electrical energy

BATTERIES AND CELLS

This energy change from chemical to electrical takes place in a cell. Sometimes it is called a voltaic cell. A cell consists of two electrodes and an electrolyte. The electrodes are usually metals and the electrolyte is a liquid or a paste. In Volta's first battery the electrodes were made of zinc and silver, and the electrolyte was salt water.

You can repeat his experiment in the science lab with a beaker of salt water and two metal electrodes such as magnesium and copper.

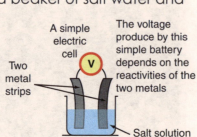

A simple electric cell

The voltage produce by this simple battery depends on the reactivities of the two metals

Two metal strips

Salt solution

The cells we buy work in the same way, but the electrolyte is a paste not a liquid. The commonest type consists of zinc and carbon electrodes, with an electrolyte of ammonium chloride. The zinc electrode forms the case of the battery.

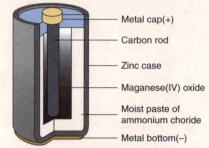

Metal cap(+)
Carbon rod
Zinc case
Maganese(IV) oxide
Moist paste of ammonium choride
Metal bottom(−)

These are cheap disposable cells with an output of about 1·5 volts. To produce a larger voltage we need to use several cells.

TOP TIP

In a battery, chemical energy is transferred to electrical energy. If the battery can be recharged the process is reversed and electrical energy is transferred to chemical energy.

46

ELECTRICITY AND WAVES

ACTIVITIES

QUESTIONS

1. What energy change takes place when a battery is being used?
2. What energy change takes place when a battery is being recharged?
3. Explain what is meant by a primary cell.
4. Explain what is meant by a secondary cell.
5. Name the three things that make up a cell.
6. What is the difference between a battery and a cell?
7. How many 1.5 cells are needed to make a 6 volt battery?

AND NOW TRY

1. Make a battery using a lemon as the electrolyte.
   You can buy a kit so that it powers a digital clock http://www.maplin.co.uk/free_uk_delivery/Lemon_Clock_Experiments.
2. Batteries often contain materials harmful to the environment. Cadmium from nickel-cadmium (or NiCad) batteries and sulphuric acid from car batteries are just two examples. They are often sent to landfill but should really be recycled so that the harmful chemicals can be removed. Find out where your nearest battery collection point is.
3. Find a dealer that sells hybrid cars. Maybe you can persuade a family member to take a test drive.

FORCES, ELECTRICITY AND WAVES

MAKE THE LINK

CDT – you learn about materials and their uses.

PSE – you learn about the impact cars can have on the environment, and alternative means of transport.

Computing – batteries play an important part in computer hardware.

DID YOU KNOW?

The earliest type of rechargeable battery is the lead acid accumulator used in our cars. It stores energy to start the engine and to operate the car's electrics. A car battery usually consists of six 2-volt cells giving an output of 12 volts. Recently, nickel cadmium (NiCad) batteries and lithium ion (Li-ion) batteries have been developed for phones, MP3 players, cameras, computers and other portable electronic devices. It is important to consider the life of the battery when choosing a phone or a computer.

47

OUR AMAZING WORLD

Primary cells

The zinc carbon cell is called a primary cell. This really means a non-rechargeable cell. As it produces electricity the materials in the electrodes are used up and cannot be replaced.

Secondary cells

Secondary cells are rechargeable. When you use them the energy change is the same as for a primary cell: chemical energy → electrical energy

But when you recharge them the reaction can be reversed so that they can be used again. When charging, the energy change is: electrical energy → chemical energy

CIRCUITS

WHAT IS A CIRCUIT?

A circuit is something you can go round. Racing cars travel round a circuit. Athletes complete a training circuit. Celebrities on tour travel on a circuit. It involves travelling round a loop and coming back to the starting point.

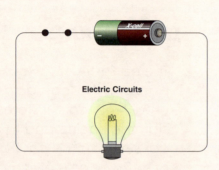

Electric Circuits

An electrical circuit is no different. The electricity must flow round the loop and return to the starting point.

If there is a gap in the circuit the electricity will not flow and the bulb will not light.

CIRCUIT SYMBOLS

So that circuits can be drawn quickly and clearly scientists and engineers use circuit symbols like those shown below.

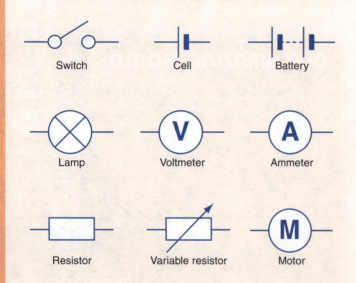

Switch	Cell	Battery
Lamp	Voltmeter	Ammeter
Resistor	Variable resistor	Motor

The circuit in the picture across would now be drawn like this.

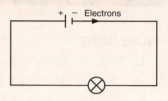

+ − Electrons

We have used the symbol for a cell though we could have used the one for a battery. As we saw in the section on batteries, most people don't distinguish between batteries and cells.

Negatively charged electrons flow round the circuit from the negative terminal to the positive terminal of the cell. The cell provides the energy, which enables the electrons to move. As the electrons flow through the bulb electrical energy is converted to heat and light.

AN ANALOGY

Electrons flow around the circuit. The trouble is, no one has ever seen an electron and no one ever will be able to see an electron. Sometimes imagining a similar situation helps us understand. Imagine the electrons are cars travelling round a road circuit and the battery is the petrol station that provides them with the energy. When we think of a similar situation like this we call it an analogy.

We will use this analogy to help us understand some of the circuit rules in the next section.

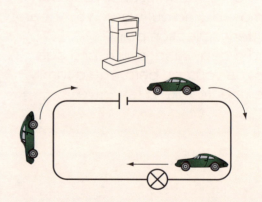

48

ELECTRICITY AND WAVES

ACTIVITIES

QUESTIONS

1. What charge do electrons have?
2. In which direction do the electrons flow round a circuit?
3. What does the battery do?
4. What energy change takes place in a bulb?
5. Draw another circuit with a battery and a bulb but this time add a switch to control the bulb.
6. Draw a circuit with a battery, a motor and a variable resistor to control the speed of the motor.
7. What energy change takes place in an electric motor?

AND NOW TRY

1. If you have a set of Christmas tree lights switch them on and try pulling one of the bulbs out. What happens?
2. The filament light bulb was invented by Thomas Edison. He invented many things in his lifetime, including the first method of recording sound.
3. Some torches and Christmas tree sets now use LEDs instead of bulbs. Find out what an LED is; when they were invented and their advantages.
4. Search for information about Benjamin Franklin and write a short biography.

TOP TIP

To allow an electric current to flow there must be a complete circuit from one terminal of the cell to the other. The cell provides the energy to push the electrons round the circuit.

MAKE THE LINK

English – you learn to use words like analogy when talking about texts.

History – you learn about inventions that changed the way we live.

CDT – you learn about the effect of technology on our lives.

History – Benjamin Franklin was one of the 'founding fathers' of the USA.

DID YOU KNOW?

Christmas tree lights are usually connected in series

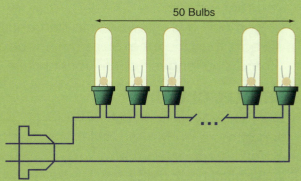

50 Bulbs

49

So if one bulb is removed or is loose they all go out or a section of them goes out. If one filament breaks, that bulb goes out but the rest stay on. This is because the bulbs contain an extra wire called a shunt. If the filament breaks, the shunt activates and keeps the current flowing.

OUR AMAZING WORLD

There are different types of lightning and scientists do not fully understand how they are formed. The most spectacular and the most dangerous is cloud to ground lightning. For a fraction of a second a current of more than 10, 000 Amps flows between the cloud and the ground and the temperature can be high enough to melt sand and turn it into glass.

CIRCUIT LAWS

When the current has to go through one bulb then another like this:

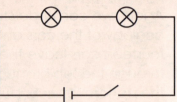

We call it a series circuit (things in a series come one after the other). There are some important things to note:

- If the switch is opened then both the bulbs go out. We can't control them independently.
- If we add a third bulb to make a series of three, the bulbs get dimmer and the current reduces.
- If one bulb is unscrewed or its filament breaks then the other bulbs go out, because there is no longer a continuous circuit.

PARALLEL CIRCUITS

Another way to connect two or more bulbs in a circuit is to connect them **in parallel**. Look at the diagram.

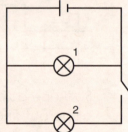

It's easy to remember. 'Series' means one after the other – think of a series of TV programmes. 'Parallel' means side by side like parallel lines. The key things about this type of circuit are:

- If we unscrew one bulb, or its filament breaks, there is still a continuous path for electricity to other bulb, so it still lights.
- You can switch off one bulb while the other one stays on. The switch in the diagram switches off bulb two but leaves bulb one lit.

You can add a third bulb or a fourth ...

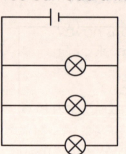

- Bulbs in our houses and cars are connected in parallel.

CURRENT

Current is the rate of flow of charge around a circuit. Using our analogy of cars going round a road circuit, current is like the number of cars passing each second. Current is measured using an ammeter and is measured in amps. In the circuit below the

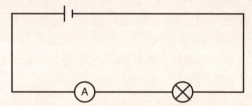

ammeter is measuring the current through the bulb.

There are two rules to know about current.

50

CIRCUIT LAWS

CURRENT IN A SERIES CIRCUIT

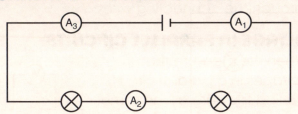

The three ammeters in this circuit measure the current in three different places. They all give the same reading. The current in a series circuit is the same at all points. We can use our cars analogy to understand this. If the cars are going round a track it doesn't matter where we count them, we will always get the same result.

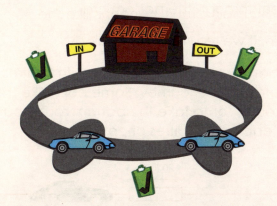

CURRENT IN A PARALLEL CIRCUIT

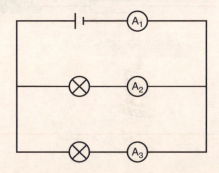

Here the circuit has two branches. Ammeter 1 measures the current leaving the battery and ammeters 2 and 3 measure the current in the branches. The currents in the two branches add together to give the current in the main circuit.

Reading on ammeter 1 = reading on ammeter 2 + reading on ammeter 3.

Again, we can understand this using the analogy of the cars and the track. Suppose fourteen cars leave the garage. Seven go through the top branch, seven go through the bottom branch, but the total number of cars is fourteen.

VOLTAGE

The voltage measures the energy available to drive the current round the circuit. If we come back to our analogy, it is like the petrol used by the cars as they travel round the track. Voltage is measured using a voltmeter and is measured in volts. In the circuit across, the voltmeter measures the voltage available to drive the current through the bulb.

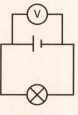

51

CIRCUIT LAWS

VOLTAGE IN A SERIES CIRCUIT

When we measure voltages in a series circuit they add up to the supply voltage.

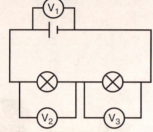

Voltmeter 1 measures the energy used to drive the current through the whole circuit.

Voltmeter 2 measures the energy used to drive the current through bulb 1.

Voltmeter 3 measures the energy used to drive the current through bulb 2.

Reading on voltmeter 1 = reading on voltmeter 2 + reading on voltmeter 3.

Coming back to our analogy:

The battery provides the energy to move the charges round the ciruit. Think of the battery voltages as the volume of petrol it provides to each car for the journey.

The battery provides each car with 20 litres of petrol for the journey. It then uses 10 litres to travel through the first bulb and 10 litres to travel through the next one.

10 litres + 10 litres = 20 litres (the total energy for the journey).

We assume no energy is needed to move through the connecting wires.

VOLTAGE IN PARALLEL CIRCUITS

When we measure the voltages in a parallel circuit we find they are all the same.

Voltmeter 1 measures the supply voltage.

Voltmeter 2 measures the energy used to drive the current through the top branch.

Voltmeter 3 measures the energy needed to drive the current through the bottom branch.

Again, if we return to our analogy:

The battery provides each car with 20 litres of petrol for the journey. Some cars take the top route, some take the bottom route, but they each use 20 litres of petrol.

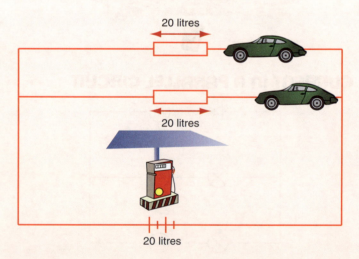

20 litres

20 litres

20 litres

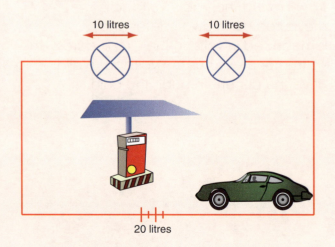

10 litres   10 litres

20 litres

52

CIRCUIT LAWS

ACTIVITIES

1. Draw a circuit with two bulbs connected in parallel and a switch which turns them both off and on together.

2. Draw a circuit with two bulbs connected in parallel and a switch which controls only one of the bulbs.

3. In the circuit below, ammeter 1 reads 0.4 A. What is the reading on ammeter 2?

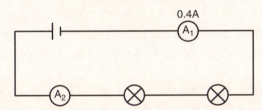

4. In the circuit below ammeter 1 reads 0.6 A and ammeter 2 reads 0.3 A. What is the reading on ammeter 3?

5. In the circuit below voltmeter 1 reads 6 V and voltmeter 2 reads 3 V. What is the reading on voltmeter 3?

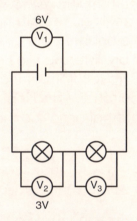

6. In the circuit below voltmeter 1 reads 6 V. What are the readings on voltmeters 2 and 3?

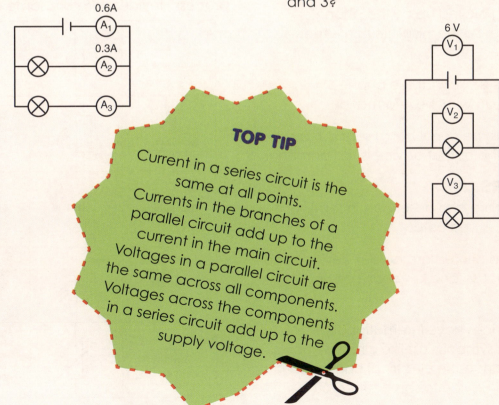

TOP TIP

Current in a series circuit is the same at all points. Currents in the branches of a parallel circuit add up to the current in the main circuit. Voltages in a parallel circuit are the same across all components. Voltages across the components in a series circuit add up to the supply voltage.

53

ELECTRONIC SYSTEMS

Electronic devices are now part of everyone's life. For work and for leisure we all depend on computers, mobile phones, MP3 players and lots of other devices. Science and technology have played an important role in developing and manufacturing these appliances. While they seem, and are, very complex they can all be understood by thinking of them as comprising of three parts: **input**, **process** and **output**.

Let's look at how this works for some everyday devices. You are having a chat on your mobile phone.
When you are listening the **input** is the radio signal that travels from a nearby phone mast to your phone and is picked up by the phone's internal aerial. So the input device is the aerial.

The **process** is what the electronics inside the phone do with the information.

Output is the sound that you can hear, so the output device is the speaker in your phone.

When you are talking, the **input** is your voice being picked up by the phone's microphone. So the input device is the microphone.

The **process** is still what the phone's electronics do with the information.

The **output** is the signal being sent out by your phone. So the output device is the antenna.

COMPUTERS

Computers are also described in terms of input process and output.

Screens and printers are output devices and scanners mouse and keyboard are input devices.

Output devices have internationally recognised symbols, just like circuit symbols. They all change the electrical energy from the process device to some other form of energy: light, sound or kinetic. Some common ones are summarised below.

Loudspeaker		Electrical to sound
Bulb		Electrical to light + Heat
Light emitting diode (led)		Electrical to light
Motor	M	Electrical to Kinetic

INPUT DEVICES

The frost warning from a car's computer must have an input device to sense the change in temperature. This could be a thermistor or a thermocouple. The lights may come on automatically when it gets dark, so here the input sensor will be a light-dependent resistor or a photocell. Input devices respond to some form of energy like light or sound and produce an electrical signal to be fed to the process device. Here are some examples.

Microphone		Converts sound to electrical energy
Light dependent resistor (LDR)		Resistance decreases with increase in light level
Thermocouple		Convers heat to electrical energy
Solar cell		Convert light to electrical energy
Thermistor		Resistance decreases with increase in temperature

ACTIVITIES

QUESTIONS

1. What are the three stages of an electronic system?
2. When you receive a text message, what is the output device?
3. When you send a text message what is the input device?
4. Which two input devices detect changes in temperature?
5. Which input device might be used so that a car's computer knows when to switch on the lights?
6. Into which form do input devices change energy?
7. Give two output devices for a computer.
8. Give two input devices for a computer.

AND NOW TRY

1. Sketch out the digits 0 to 9 as they would be shown on a seven-segment display.
2. Find out which models of cars have LED indicators.
3. Which town has converted all its street lighting to LEDs?
4. How much money have some cities saved by changing to LED traffic lights?
5. Street lights turn on automatically when it gets dark. Can you see the sensor that lets them do this?
6. For another cool use for LEDs and photovoltaic cells used together have a look at GreenPix: http://www.greenpix.org/.

MAKE THE LINK

Computing – you learn about input and output devices.

Drama – you learn about stage lighting. LED stage lights mean the actors don't get so hot under the lights.

DID YOU KNOW?

A normal filament bulb turns most of the electrical energy supplied into heat and only about 10% of it into light. Light-emitting diodes (LEDs) turn nearly all the energy into light. They don't get hot in normal use and are sometimes called cold light sources.
A group of seven LEDs can be combined to make a seven-segment display.

http://wpcontent.answers.com/wikipedia/
By switching on the different sections it is possible to display the digits 0–9.

55

OUR AMAZING WORLD

LEDs only emit a single colour of light but by using a cluster of red, green and blue leds it is possible to create the effect of white light (see the section on colour on page 59). We can now buy LED torches and lamps. Because they use less energy and don't produce heat, LEDs are now being used in all sorts of ways. Some cars use them as indicators, they are common in traffic lights and in some places they are even used in street lighting.

TOP TIP

Electronic systems are described by the three stages input, process and output. Output devices change electrical energy into some other form like light or sound. Input devices convert other forms of energy to electrical signals.

SOUND

Sound is a form of energy and travels in waves.

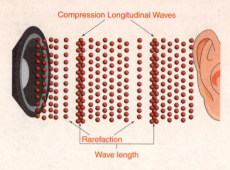

Compression Longitudinal Waves

Rarefaction

Wave length

The speaker vibrates backwards and forwards. Sound is called a **longitudinal wave**, which means the waves travel in the same direction as the vibrations. When the speaker moves forwards it produces areas of higher pressure called **compressions**. When it moves back it produces areas of lower pressure called **rarefactions**. These pressure variations travel through the air and are detected by the listener's ears.

SOUND TRAVELS

Sound waves can travel through solids, liquids and gases. They cannot travel through a vacuum but need to travel through some medium, like air or water. In the diagram when the air is pumped out of the bell jar the sound no longer reaches us.

PITCH AND FREQUENCY

All sound is produced by something vibrating. In the picture above, the cone of the speaker is vibrating. If it vibrates 100 times per second we say the frequency is 100 hertz (Hz) and if it vibrates 1 000 times per second we say the frequency is 1 000 Hz or 1 kHz.

'Pitch' and 'frequency' really mean the same thing. Frequency is a scientific term, while pitch is a musical term. A scientist might describe a sound as high frequency while a musician would call it high pitched.

RANGE OF HEARING

All people vary but as a general rule people can hear sounds when the frequency is

above 20 Hz and below 20 kHz. Vibrations with a frequency above 20 kHz are called ultrasound and a number of animals can hear these. Dolphins and whales are thought to communicate at these high frequencies, and bats use ultrasound to find insects for food.

The chart below shows the hearing range of some animals.

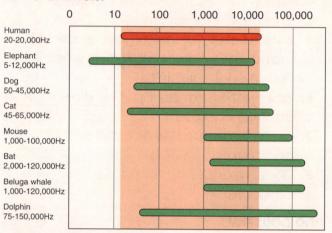

SPEED OF SOUND

Sound travels at a speed of about 340 metres per second. This is about 700 miles per hour – a little faster than a jet airliner. While this seems fast, it is very much slower than the speed of light. In one second sound can travel about three times the length of a football pitch – light, on the other hand, travels seven times round the earth in the same time. There are many cases in which we notice the difference between the speeds of light and sound. In a thunder storm the thunder and the lightning happen at the same time but we see the lightning before we hear the thunder. When watching fireworks we see the firework explode before we hear the bang.

TOP TIP
The speed of sound is 340m/s, which is much less than the speed of light.

56

ELECTRICITY AND WAVES

ACTIVITIES

QUESTIONS

1. What name is given to the areas of higher pressure in a sound wave?
2. What are the highest and lowest frequencies that people can hear?
3. What name is given to vibrations that are too high frequency for people to hear?
4. Name two animals that can hear frequencies above 100 kHz.
5. Name an animal that can hear frequencies below 20 Hz.
6. What is the speed of sound?
7. In a storm, the lightning is seen three seconds before the thunder is heard. How far away was the lightning flash?

AND NOW TRY

1. In a thunderstorm count the seconds between the lightning flash and the thunder. Calculate how far you are from the thunder.
2. 'Silent' dog whistles actually produce sound above the range of human hearing. If you have a pet, try to see if it responds to one.
3. Find a spot where you hear a good echo and try to estimate the speed of sound.
4. Find out what is meant by infrasound and how it can be used.

MAKE THE LINK

Music – you learn about notes and musical scales.

History – you learn how echoes were used to find submarines in World War Two.

Maths – you learn to calculate using formulae like speed = distance/time accurately.

DID YOU KNOW?

We can use frequency to understand the notes in music. Middle C has a frequency of 256 Hz.
If we move up an octave to High C then the frequency doubles to 512 Hz. It works for all notes in all scales. Going up the scale by one octave doubles the frequency.

57

OUR AMAZING WORLD

For over a century X-rays have been used to form pictures of the inside of our bodies. X-rays are very good for detecting broken bones but they don't detect soft tissue like muscles and internal organs very well. Also, X-rays can be harmful to human tissues so doctors are reluctant to X-ray unborn babies. Instead, an image of the baby in its mother's uterus is formed using ultrasound.

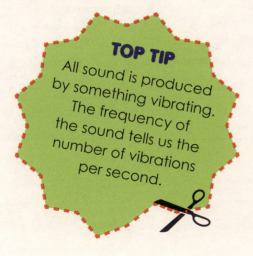

TOP TIP
All sound is produced by something vibrating. The frequency of the sound tells us the number of vibrations per second.

REFLECTION AND REFRACTION

Light is a form of energy. It travels in waves. The speed of light is 300 000 000 m/s. This means that light can reach us from the moon in a bit over one second, and from the sun in eight minutes. We see things when light is reflected from them into our eyes.

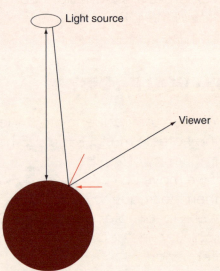

Light source

Viewer

The normal is a line drawn at right angles to the surface of the glass. The angles are always measured between the rays and the **normal**.

When you look in the mirror the rays of light from your face are reflected back into our eyes. They seem to come from behind the mirror and you see an image (your reflection).

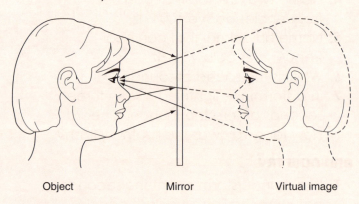

Object   Mirror   Virtual image

REFLECTION

When light strikes a flat mirror it is reflected at the same angle it arrived at. We represent this in an optical diagram like the one below and say that:

angle of incidence = angle of reflection

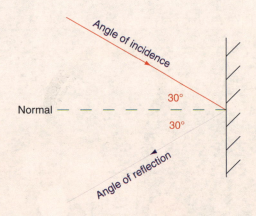

Angle of incidence

Normal

30°

30°

Angle of reflection

REFRACTION

When light travels from air to glass it slows down and changes direction. When it comes back out into the air it returns to its original speed and direction.

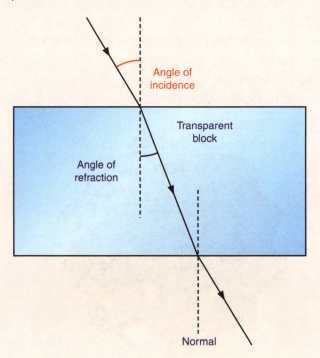

Angle of incidence

Transparent block

Angle of refraction

Normal

REFLECTION AND REFRACTION

Notice again how the angles are measured to the normal, not the glass surface.

This effect is called **refraction**. Light is also refracted by water, clear plastic and all transparent materials but for the rest of this section we will talk about glass. The refraction of light can produce beautiful effects like rainbows. By understanding refraction it is also possible to design lenses for cameras, spectacles and lots of other applications.

SPECTRA

White light consists of a mixture of colours called a spectrum (the plural of spectrum is spectra). The colours all have different wavelengths and when the light enters glass some colours are refracted more than others.

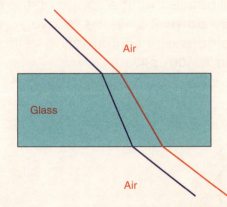

The diagram shows that blue light is refracted more than red light. By passing the light through a triangular piece of glass called a prism this effect can be increased and a spectrum produced:

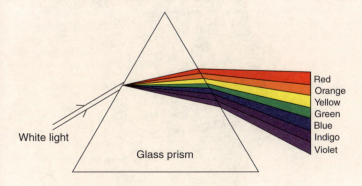

LENSES

A lens is a circular piece of glass. The glass at the centre is either thicker or thinner than at the edges. There are two types of lenses: **convex**, which are thicker in the middle and **concave** which are thinner in the middle. Convex lenses make the light *converge*. They focus the light. Concave lenses make the light *diverge*.

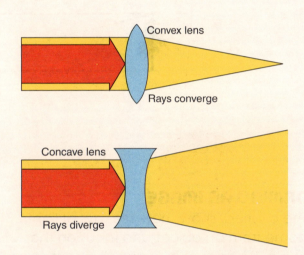

Both of these lens types are used in spectacles. We will look at two uses of convex lenses.

MAGNIFYING GLASS

If we look through a convex lens at something close up, the lens acts as a magnifying glass.

The lens bends the rays of light in such a way as to make the print seem bigger than it really is.

59

REFLECTION AND REFRACTION

DISTANT OBJECT

When light from a more distant object falls on a convex lens the lens focuses the light and forms an image as shown in the diagram.

The lens in a camera does exactly the same. An image is formed on the film or the electronic sensor inside the camera and recorded. Notice that the image is upside down.

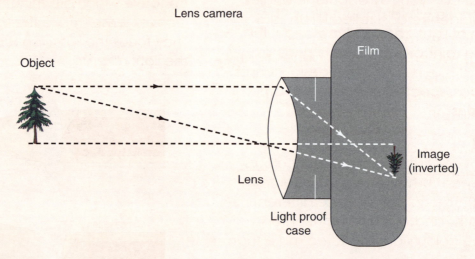

Lens camera

Object

Film

Lens

Image (inverted)

Light proof case

FORMING AN IMAGE

The overleaf shows how the rays of light come from an object and focus on the retina. It happens in exactly the same way as in a camera.

You will notice that the image on the retina is **upside down.** It's like this for everyone – the brain processes the information so that you see things correctly.

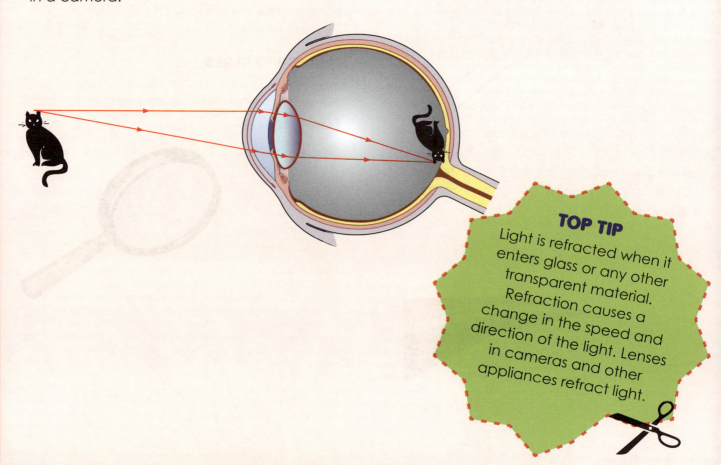

TOP TIP
Light is refracted when it enters glass or any other transparent material. Refraction causes a change in the speed and direction of the light. Lenses in cameras and other appliances refract light.

60

ELECTRICITY AND WAVES

ACTIVITIES

QUESTIONS

1. What name is given to the line at right angles to the surface of glass or a mirror?
2. What can you say about the angles of incidence and reflection in light hitting a mirror?
3. What two changes occur when light is refracted?
4. Which is refracted more, red light or blue light?
5. What name is given to the colours produced when white light travels through glass?
6. Name the two types of lens.
7. What happens when you look at something close up through a convex lens?
8. Give another use of a convex lens.

AND NOW TRY

1. A mnemonic is a trick to help you remember things easily. Do you know or can you find a mnemonic for the colours of the spectrum?
2. Sometimes you see a double rainbow. Can you find a picture to explain how it is formed?
3. You can make a water drop magnifier have a look at http://www. thenakedscientists.com/ for some great ideas to try at home.
4. Short-sighted people wear glasses with concave lenses and long-sighted people wear convex lenses. You can tell the difference by looking at their glasses because convex lenses magnify things while concave lenses make them look smaller.

FORCES, ELECTRICITY AND WAVES

MAKE THE LINK

Art – you learn about colours.

Drama – different-coloured stage lights are used to produce dramatic effects.

English – you learn how nouns like spectrum take their plural.

OUR AMAZING WORLD

A rainbow is a special type of spectrum. When you are looking at a rainbow the sun is behind you and shining on the raindrops.

The light is reflected back towards you inside the raindrop and refraction splits the white light into a spectrum as shown above.

ELECTROMAGNETIC SPECTRUM

We saw in the last section that white light is a mixture of colours. These colours all have different wavelengths and can be split up into a spectrum. We call this the visible spectrum because it is the range of wavelengths that people can see. Visible light is in fact just a small part of a much larger family of waves called the electromagnetic spectrum as shown below.

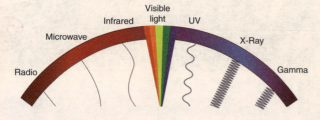

Radio waves have the longest wavelengths and gamma rays the shortest. All these waves, including light, travel at 300 000 000 metres per second (m/s). and they can travel through a vacuum, which means that they can travel through space. While humans can only see the small section in the middle that we call the visible spectrum, some animals have adapted to see infrared and ultraviolet.

RADIO WAVES

Radio waves have the longest wavelengths of all the electromagnetic spectrum. The longest wavelengths, up to 1 000 m, carry information to radios in our cars and homes. Shorter wavelengths of a few metres bring us television while mobile phones use wavelengths of about 15 cm.

MICROWAVES

Microwaves have a wavelength of about 3 cm. They are now part of our daily lives and we use them in microwave ovens to heat our food. The first application of microwaves was radar, their use for cooking did not come until much later. Modern radars still send out microwaves.

The microwaves are reflected back from the target and allow us to track aircraft, thunderstorms and lots of other things.

INFRARED

Anything that is hotter than its surroundings gives off infrared radiation. It is sometimes called radiant heat. The wavelength of these waves is less than 1 mm and while we can't see them, special infrared cameras can detect the heat coming from warm objects.

ULTRAVIOLET

Ultraviolet waves have a wavelength less than visible light. These and the next two members of the electromagnetic spectrum have very small wavelengths. Ultraviolet is present in the sun's rays and causes sunburn. Some materials glow when ultraviolet waves strike them. This effect called fluorescence is used to put security markings on banknotes.

X-RAYS

X-rays have very high energy and can penetrate the human body. Wilhelm Rontgen found how to take X-ray photographs by shining X-rays through the body onto a special film.

X-rays can be harmful to healthy cells so they are only used if really necessary and are very rarely used on unborn babies.

GAMMA RAYS

Gamma rays are produced by certain radioactive materials like some forms of cobalt. They have very high energy and are harmful to living cells. They can be used in medicine in two ways. A gamma camera can be used to give information about some organs of the body, like the liver or the lungs. They can also be used to kill certain cancer cells.

62

ACTIVITIES

QUESTIONS

1. What is the speed of all members of the electromagnetic spectrum?
2. Which type of wave in the electromagnetic spectrum has the shortest wavelength?
3. What is another name for infrared radiation?
4. X-rays are rarely used on unborn babies. What is used to produce a picture of an unborn baby?
5. What effect does ultraviolet radiation have on humans?
6. Who discovered X-rays?
7. Where do gamma rays come from?
8. What is a nanometre?
9. What was the first application of microwaves?
10. Give a use for gamma rays.

AND NOW TRY

1. Remote controls for TVs and other appliances send an infrared signal. You can't see it but try pressing the buttons and watching with your camera phone.
2. You can buy a security pen and UV light for a few pounds at Amazon. Try using it to put secret markings on your property.
3. The global hydrology and climate centre allows you to look at infrared photographs of the earth. Have a look at their website: http://www.ghcc.msfc.nasa.gov/GOES/globalir.html

MAKE THE LINK

History – the first practical radar was developed in 1935 and by 1939 Britain had a chain of radar stations along the south of England. The fact the Allies were ahead in the development of radar was an important factor in World War Two.

English – 'infrared' means below red and 'ultraviolet' means beyond violet. You can learn to use these prefixes with other words in English.

DID YOU KNOW?

Some snakes have infrared sensors as well as eyes. This means they can 'see' warm-blooded prey in the dark.

OUR AMAZING WORLD

63

While humans can't see ultraviolet it is visible to some insects. The two pictures below show how an evening primrose looks to a human on the left and how it might look to a bee on the right. The bee can see ultraviolet, so things look completely different.

TOP TIP

Visible light, which humans can see, is part of a family of waves called the electromagnetic spectrum. All these waves travel through space at a speed of 300 000 000 metres per second. They have different wavelengths and a wide variety of applications.

EYES AND EARS

In the last two sections we have looked at light waves and sound waves. We will now look at how our eyes and ears have adapted to receive these waves and pass the information to the brain.

EYES

In the section on refraction we saw how a convex lens forms an image on retina of the eye.

The parts of the eye and the jobs they do are shown in the table below.

EARS

In the section on sound we saw vibrations travel as waves through the air. The sound waves are collected by the flap of your outer ear and cause the eardrum to vibrate.

The parts of the ear and the jobs they do are shown in the table below.

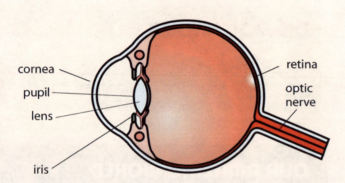

cornea
pupil
lens
iris
retina
optic nerve

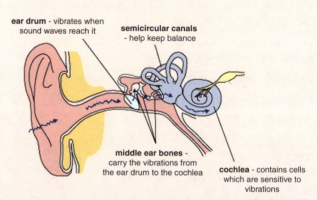

ear drum - vibrates when sound waves reach it

semicircular canals - help keep balance

middle ear bones - carry the vibrations from the ear drum to the cochlea

cochlea - contains cells which are sensitive to vibrations

Parts of eye	Purpose
Pupil	... where the light enters the eye
Lens and cornea	... act together to focus the light on the retina. Tiny muscles change the shape of the lens to suit near and distant vision.
Retina	... contains special light-sensitive cells which produce tiny electrical signals to be passed to the brain
Iris	... is the coloured part of the eye. It moves to control the amount of light entering your eye.
Sclera or sclerotic coat	... is the tough white part of the eye
Optic nerve	... carries the signals from the retina to the brain

Parts of ear	What they do
Ear drum	This is a thin layer of skin. Sound waves reaching the eardrum cause it to vibrate.
Hammer, anvil and stirrup	These are three tiny bones. They are in fact the smallest bones in the human body. When the eardrum vibrates it causes these small bones to vibrate. They pass the vibrations to the cochlea.
Cochlea	This is filled with watery fluid. When this fluid vibrates nerve cells produce electrical signals to pass to the brain.

ELECTRICITY AND WAVES

ACTIVITIES

QUESTIONS

1. Where is the image formed in the eye?
2. What name is given to the coloured part of the eye?
3. What does the lens of the eye do?
4. How is the information passed from the retina to the brain?
5. What energy change takes place in the retina?
6. Name the three small bones in the ear.
7. In which part of the ear are the vibrations changed to electrical signals?
8. What is the purpose of the outer ear flap?

AND NOW TRY

1. Each of our eyes has an area with no light receptors. This is where the optic nerve meets the eye and it is known as your blind spot. You are not normally aware of it but you can show it's there with this simple test:

Stare at the star in picture above. Now cover your left eye. Move the page slowly towards you, always concentrating on the star. At some point the dot will disappear. This is because the light from the dot is now falling on your blind spot.

2. Having two ears lets us judge the position of sound. Can you devise an experiment in which you blindfold your partner to see if they can judge where sound is coming from?

FORCES, ELECTRICITY AND WAVES

MAKE THE LINK

Modern Studies – you learn about the elderly and how they need help as their hearing and eyesight deteriorate.

English – the senses, such as sight and hearing, are often a good stimulus for writing creatively.

Art and Design – you learn about different ways of 'seeing' a picture.

Music – you develop your listening skills.

DID YOU KNOW?

The outer flap of the human ear has developed to collect sound waves and pass them to the eardrum. Some animals have much more highly developed outer ears. A cat's ears collect the waves in the same way as a satellite dish collects radio waves. Cats and some other animals can turn their ears around to collect more sound.

65

OUR AMAZING WORLD

The lens in the eye changes shape depending on the position of the object you are looking at.

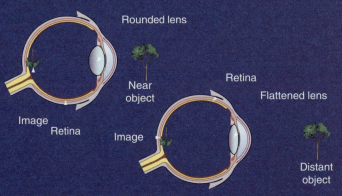

Rounded lens

Near object

Retina

Flattened lens

Image

Retina

Image

Distant object

Notice how the lens is much thicker for a near object. When people get older their eye lenses cannot change as easily and they need glasses for reading.

SUMMARY ACTIVITIES

TRANSPORT SYSTEMS

Car and road systems are a vital part of our economy. We depend on them for business and leisure but they have a huge impact on the environment and consume large amounts of fossil fuels. This project links Business Studies, Maths and ICT to look at the transport and communications issues involved in working with people a long way away.

1. Let's imagine we are working on a design project and we have a partner in London. Work in pairs and make some drawings of anything you like: an item of clothing, a car, a piece of sports equipment, a building, etc. When you have finished, imagine that we want our partners in London to look at your design and give us feedback. How will we go about this? It would be nice to travel to London and have a meeting with our colleagues. There are three main options: we could drive, fly or travel by train. We will consider four things.

COST

- How much will it cost for two people to fly to London? Remember to include taxi fares to and from the airport.
- How much will it cost to travel by train? Again, remember that you have to get to the station.
- We could hire a car and drive there and back. Cost this one out. Remember the petrol!

TIME

- Work out how long each journey will take. Remember to include the time it takes to get to the airport or the station, check-in times and the time to get from the airport to the centre of London.

SAFETY

- All businesses are responsible for the safety of their employees. How do the journeys rate in terms of human safety?

ENVIRONMENTAL IMPACT

- As responsible citizens we should consider the environmental impact of our journeys, so let's look at the carbon dioxide produced by each of the options.
- There are many carbon footprint calculators available on the internet. Make a table to summarise your findings.

You might decide that you don't really need to make the journey and that you can share your ideas using ICT. The first step would be to take your design and scan it. Save the scanned image either as a jpeg or a PDF file. Now attach it to an email and send it to another group in your class. Let's try two things:

- Have your classmates email you some feedback and use this to modify your design.
- Set up a video conference to discuss your ideas. You might be able to do this on a school network or you can use Skype http://www.skype.com/intl/en-gb/.

Alternatively, there are companies that specialise in video conferencing. Try searching for some options and find out how much they cost. Try http://www.megameeting.co.uk/.

ACTIVITIES

1. Match the following key words with their meanings below.

Keywords:
- a) friction
- b) field force
- c) density
- d) circuit
- e) cell
- f) series circuit
- g) frequency
- h) refraction
- i) microwaves
- j) convex lens

Meanings:
1. two electrodes and an electrolyte used to change chemical energy into electrical energy
2. the bending of light when it enters glass or water (or any other transparent material)
3. a force between surfaces that opposes motion
4. a lens which curves outwards
5. a circuit where one bulb comes after the other
6. electromagnetic waves of wavelength approximately 3 cm
7. the mass of 1 cm³ of a material
8. the number of waves per second
9. a force felt when two objects do not touch each other
10. a loop around which charges can flow

2. An audiometrist collects data on the highest frequency that twenty different patients can hear. Here are her results.

Calculate the average value for the highest frequency the patients could hear.
Analyse the results into ranges in a table like this. Use this data to draw a histogram.

Patient	Frequency (kHz)
1	15
2	16·5
3	18
4	17
5	17·5
6	14·5
7	19
8	18·5
9	19·5
10	20
11	16
12	17
13	18
14	18·5
15	17·5
16	15·5
17	16
18	17
19	18
20	17·5

Range(kHz)	No. of patients
12·5–14	
14·5–16	
16·5–18	
18·5–20	

MINDMAP

Complete the following mindmap.

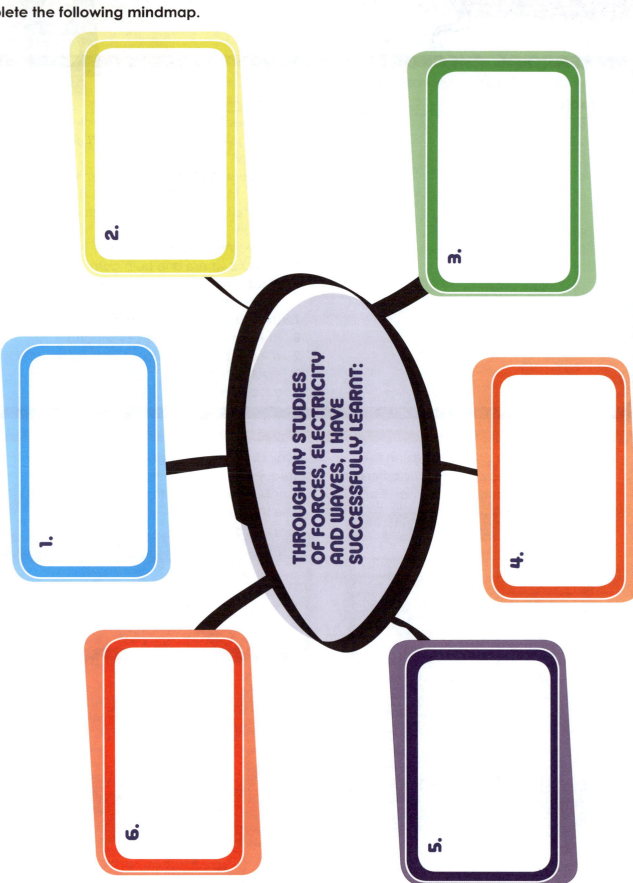

1.

2.

3.

4.

5.

6.

THROUGH MY STUDIES OF FORCES, ELECTRICITY AND WAVES, I HAVE SUCCESSFULLY LEARNT:

68

MY LEARNING CHECKLIST

	Help needed	Getting there	Good to go!
1. I have investigated the effects of friction and can identify situations where friction is necessary and where it should be reduced.	◯	◯	◯
2. I can distinguish between mass and weight and can predict how the weight of an object changes in different situations.	◯	◯	◯
3. I can calculate density and explain floating and sinking in terms of density.	◯	◯	◯
4. I can distinguish between series and parallel circuits and understand their applications in everyday life.	◯	◯	◯
5. I can explain electronic systems in terms of input, process and output.	◯	◯	◯
6. I can explain the energy changes in cells and understand the importance of this technology to society.	◯	◯	◯
7. I have investigated the refraction of light and can use this to explain the action of lenses and prisms.	◯	◯	◯

69

BREAK IT DOWN

When dealing with anything complex it is often helpful to break it down into smaller parts. Think of your house like this. It has:

- a heating system to keep it warm
- a water system to supply you with fresh water
- a sewage system to take away the waste products

This way of thinking about complex situations also works for the human body. Our bodies are made up of about a hundred million million cells and we can think of them as being grouped into systems. Each system has a job to do. Here are some of the main ones:

System	Purpose
Circulatory system	moves blood around the body
Respiratory system	takes oxygen from the air and transfers it to our blood, then takes carbon dioxide from our blood and puts it back in the air
Digestive system	helps us get the energy and nutrients out of food and into our body
Immune system	Defends us against bacteria and viruses and other things. This is explained in the section on defence that follows this one.
Reproductive system	Allows us to have children. This is covered in the section on fertilisation (page 82)
Sensory system	Helps us detect what is going on around us. It involves eyes and ears, which are explained in section on eyes and ears (page 64). Senses also include touch, smell and taste.

CIRCULATORY SYSTEM

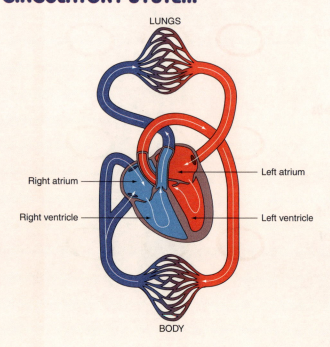

LUNGS

Right atrium

Left atrium

Right ventricle

Left ventricle

BODY

Your circulatory system carries blood around your body. Blood is pumped round by your heart and carries oxygen, food, heat and dissolved chemicals to all the cells of the body. The diagram shows a simplified form of the route taken by the blood.

The right-hand side of the heart pumps blood to the lungs, where it picks up oxygen. The left-hand side pumps the blood all round the body so that it can deliver the oxygen to the cells. In the cells the blood picks up waste products, including carbon dioxide. The carbon dioxide is then carried back to the lungs, where it is breathed out.

70

BREAK IT DOWN

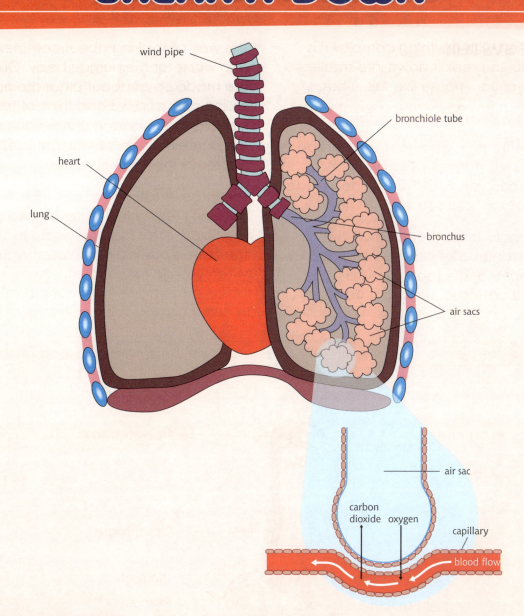

wind pipe

bronchiole tube

heart

lung

bronchus

air sacs

air sac

carbon dioxide   oxygen

capillary

blood flow

RESPIRATORY SYSTEM

When you breathe in your ribcage moves up and outwards and your diaphragm moves down. Air enters your body via the mouth and nose, passes down your trachea and into the bronchial tubes. In the lungs the bronchial tubes branch out like the twigs of a tree and finally end in little air sacs. In the air sacs oxygen from the air passes into tiny blood vessels called capillaries, where it is picked up by red blood cells. Waste carbon dioxide is passed in the opposite direction from the blood into the air sac so that it can be breathed out.

TOP TIP

Anything complex, whether it is the human body or a computer, can be understood better by thinking of it in terms of systems.

BREAK IT DOWN

DIGESTIVE SYSTEM

Your digestive system takes big lumps of food and breaks these down to small molecules of food which are easily absorbed. It starts when you chew your food. The chemicals in saliva mix with the food and start to break it down into substances which your body can use. When you swallow, the food passes down the oesophagus to your stomach.

In the stomach, acids break down the food into even smaller particles and also kill harmful bacteria. The food is now a liquid mixture and passes to the small intestine.

The small intestine is a tube nearly 7 m long and the food takes about four hours to pass through it. It is here that the nutrients pass into the blood. The pancreas and the gall bladder produce chemicals to help this process. The material that is left behind passes to the large intestine.

The large intestine is about 1.5 m long and here any remaining water is absorbed by the body. After that the waste material – the faeces – is pushed out through your anus when you go to the toilet.

72

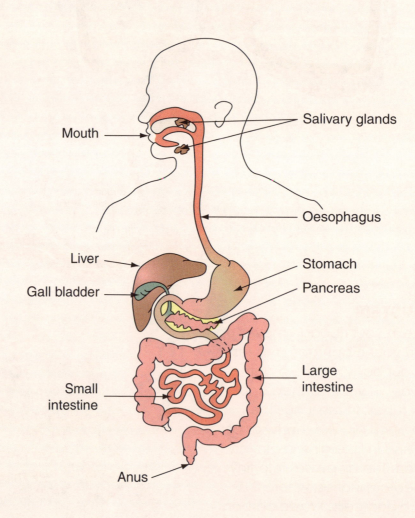

ACTIVITIES

QUESTIONS

1. What does blood carry around our bodies?
2. What is the job of the heart?
3. What happens to blood in the lungs?
4. What passes out of blood into the air sacs?
5. Which two parts of the body move so that we can breathe?
6. What carries food to the stomach?
7. What is absorbed into the blood in the small intestine?
8. What is absorbed into the body in the large intestine?
9. Where does waste material pass out of the body?

AND NOW TRY

1. Measure your heart rate by taking your pulse. Now try some vigorous exercise like running up the stairs and take your pulse again. How long does it take to return to normal?
2. Try the same activity as above but this time measure your breathing rate before and after exercise.
3. We have mentioned six body systems but there are others, like the nervous system. Find information about other body systems. The BBC website http://www.bbc.co.uk/science/humanbody/ is a good place to start looking.
4. You can measure the volume of air your lungs hold with some simple experiments. Have a look at http://www.biologycorner.com/worksheets/lungcapacity.html.

BODY SYSTEMS

MAKE THE LINK

PE – you learn about heart rate and fitness.

Home Economics – you learn about the nutrients in food.

Social Education – you learn about smoking and its effect on the respiratory system.

English – writers and poets think of the heart as the centre of emotion.

DID YOU KNOW?

Your heart beats around 100 000 times a day. Over a lifetime, the average human heart beats two and half billion times.

OUR AMAZING WORLD

Arteries carry blood away from the heart. They have thick walls and swell slightly with each beat of the heart. This swelling of the arteries with the heartbeat is your pulse, and you can feel it at your wrist and in your neck.

By counting the pulse beats it is possible to find out how many times your heart beats each minute.

Veins carry blood back to the heart they have much thinner walls than arteries and there is no pulse in veins. Veins have valves to stop the blood flowing backwards.

TECHNOLOGY AND HEALTH

HISTORY

For a long time people have realised the value of using science and technology to detect and treat illness. If you say you feel unwell the first check is usually your body temperature. If your temperature is above 37°C then we know something is wrong.

The next thing a doctor is likely to do is listen to your breathing with a stethoscope. Stethoscopes were first used at the beginning of the nineteenth century and are designed to listen to the sounds produced inside a human or animal body. They can be used to listen to the beating of the heart; air moving through the lungs; blood moving through arteries; and food moving through the intestine. These are early examples of how technology is used to detect illness. In the section on the electromagnetic spectrum (page 62) we saw how X-rays can be used to detect broken bones, and in the section on sound (page 56) we saw how ultrasound can be used to produce images of unborn babies. When these advances were combined with the processing power of modern computers, healthcare changed dramatically.

ELECTROMAGNETIC SPECTRUM

The waves of the electromagnetic spectrum described previously have many applications in healthcare.

- Infrared pictures can detect cancers. IR waves can also be used to aid the healing of muscle injuries.
- Ultraviolet waves can be used to treat some skin conditions like acne.
- X-rays can be used to make images of the inside of the body.
- Gamma rays can be used to kill cancer cells.

CT SCANS

We saw how X-rays are used to produce an image of bones in the body. Computed tomography (CT) combines the uses of X-rays and computer processing. Many thousands of X-rays are taken at different angles and a computer uses the information to build up a three-dimensional picture of the body.

The pictures below show CT scans of a human brain. The first one includes the eyes.

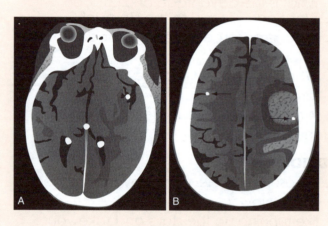

These scans give information about internal structures which would not be seen on a standard X-ray. Magnetic resonance imaging (MRI) is a more recent development and gives even more detailed images.

OPTICAL FIBRES

An optical fibre is a long thin piece of glass Light travels along inside the fibre by a process called total internal reflection:

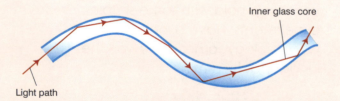

Inner glass core

Light path

Optical fibres are used in communication and in medicine. A bundle of optical fibres called an endoscope can be used to examine a patient's lungs or stomach.

74

ACTIVITIES

QUESTIONS

1. Give an example of a medical use of the following members of the electromagnetic spectrum:
   a) gamma Rays
   b) X-rays
   c) infrared
   d) ultraviolet

2. What is the advantage of a CT scan over a normal X-ray?
3. What is an optical fibre?
4. What is the purpose of a stethoscope?
5. What is normal human body temperature?
6. Give an advantage of using a laser to perform operations.
7. Which instrument uses optical fibres to see inside the body?

AND NOW TRY

1. Laser surgery is now often advertised to correct short-sightedness. Find an advert in a magazine or on the internet and make some enquiries.
   a) What is involved?
   b) How much does it cost?
   c) Is it recommended for everyone?
   d) What is the view of an optician?

2. Carry out an internet search for images produced by:
   a) CT scan
   b) MRI scan
   c) endoscope
   d) camera pill

3. Find out what Joseph Lister contributed to medical science.
4. The Hippocratic Oath is traditionally taken by doctors. What is the Hippocratic Oath and who was Hippocrates?

MAKE THE LINK

Modern Studies – you learn about healthcare and the National Health Service.

Art – optical fibres can be used in sculptures.

History – you learn about early medicine and surgery in battles and on naval ships.

DID YOU KNOW?

A pill endoscope is a camera in a pill. The patient swallows it in the usual way and as it passes down the oesophagus into the stomach and then the intestine it transmits thousands of pictures to a receiver on the patient's belt. Eventually it passes out naturally with the faeces when they go to the toilet.

OUR AMAZING WORLD

Traditionally doctors performed operations with a very sharp knife called a scalpel. Now it is possible to make incisions with a laser. A laser is a very intense beam of light; all the energy is focused on a very small area and this means it can be used instead of a scalpel to cut tissue. In laser surgery the blood vessels are sealed by the heat of the laser so the patient loses less blood.

TOP TIP

Doctors and healthcare professionals have always used technology to detect and treat illness. This ranges from the very simple step of taking someone's temperature to the use of powerful computers to build up pictures of the inside of the body.

CELLS

All living things are made of cells. Most of the organisms in the world such as bacteria and tiny plants called algae are made up of only one cell. They are said to be unicellular, and we will look at them in the section on micro-organisms (see pages 78). Larger organisms are multicellular and are made of a large variety of cell types. It is estimated that an adult human is made up of more than ten million million cells. Cells vary in shape and size because of the many different jobs they have to do. Here are some examples:

Type of cell	Purpose
Muscle cells	enable us to move
Red blood cells	carry oxygen to all the other cells in the body
White blood cells	fight infection
Nerve cells	carry messages to and from the brain
Fat cells	store energy
Sex cells	enable us to have children

Even though they do different jobs, all cells have some things in common.

A SIMPLE ANIMAL CELL

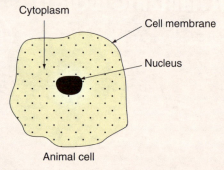

Cytoplasm
Cell membrane
Nucleus
Animal cell

The **nucleus** contains genetic material and controls the activity of the cell. Chromosomes are found in the nucleus of the cell and contain all the information about the organism. See the section on DNA, page 84, for more information on chromosomes.

The **cell membrane** controls the movement of substances in and out of the cell. It allows food to enter and waste products to leave.

The **cytoplasm** is where the chemical activities of the cell take place. It contains water and dissolved chemicals.

A CELL FROM A PLANT LEAF

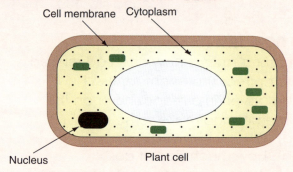

Cell membrane   Cytoplasm
Nucleus
Plant cell

In addition to the nucleus, cell membrane and cytoplasm, most plant cells have three other structures:

The **cell wall** is made of the chemical cellulose and it strengthens and supports the cell. Water and dissolved chemicals can pass through the cell wall.

The **vacuole** is a store of water in the centre of the cell. This helps to maintain the shape of the cell.

Plant leaves make food by photosynthesis so they contain food factories called **chloroplasts**. These contain the chemical chlorophyll, which absorbs the light energy needed for photosynthesis.

76

BODY SYSTEMS

ACTIVITIES

QUESTIONS

1. Which parts are found in all cells?
2. Which parts are found only in plant cells?
3. Explain the job of each of these cell parts:
   a) nucleus
   b) cell membrane
   c) cytoplasm
4. Which two structures help maintain the shape of a plant cell?

AND NOW TRY

1. Some web sites give a very good summary of the information about cells. Try http://lgfl.skoool.co.uk/keystage3.aspx?id=63.
2. Plan a trip to the National Scottish Museum to see the remains of Dolly.
3. In asexual reproduction organisms make clones of themselves. Find out what is meant by asexual reproduction.

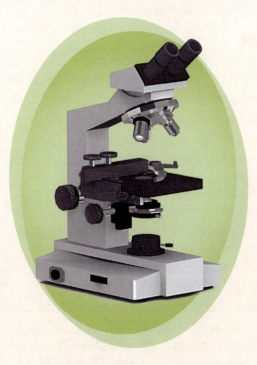

MAKE THE LINK

English – you learn to use the word 'cell' in different contexts. A cell is a unit. In a spy story you can have a cell of secret agents. **CDT –** you learn about strength of structures. The cell wall and vacuole in a plant cell strengthen the structure. **Art –** early scientists did not have cameras and became very skilled at drawing and painting what they saw with their microscopes. **ICT –** the term 'cloning' is used to describe making copies of software or hardware.

DID YOU KNOW?

Cells are very small and we can only see them using a microscope. We say they are microscopic. There are many different types of microscope but they all have some things in common. Microscopes use a system of lenses. You look through the eyepiece and the objective lens is close to the specimen. Selecting a different objective lens usually changes the magnification of the microscope. You always start out with the lowest-power objective lens.

77

OUR AMAZING WORLD

The nucleus of an animal cell contains all the genetic information for that animal's body. Cloning is where the information from one cell is used to make another animal which is identical to the first. The most famous example is Dolly the sheep, who was cloned in 1996. She lived till 2003 and her remains can be seen in the National Museum of Scotland.

TOP TIP

All cells have a nucleus, cell membrane and cytoplasm. Most plant cells also have a cell wall, vacuole and often chloroplasts.

MICRO-ORGANISMS

A micro-organism is a living thing which can only be seen with a microscope. Micro-organisms include bacteria, fungi, some microscopic plants, protozoa and plankton. Plankton is a vital part of the life system of our oceans. Micro-organisms are usually single celled but some are multicellular.

BACTERIA

Bacteria is the plural of bacterium. Bacteria are unicellular organisms and are found literally everywhere on earth. Even in radioactive waste and acidic hot springs where you would expect no life you can still find bacteria. In a gram of soil or a cubic centimetre of water there are millions of bacteria. There are also millions of bacteria on every square centimetre of our skin. When we think of bacteria we tend to think of things that make us ill but most are harmless and some are beneficial. They come in many shapes, but the commonest are spherical ones called cocci, and bacilli, which are rod shaped.

BACTERIA IN THE SOIL

Bacteria are an essential part of the life cycle of the planet. In the section on chemicals and agriculture we saw how bacteria in the roots of some plants form compounds with nitrogen from the atmosphere. We also saw how bacteria break down animal waste together with the remains of dead plants and animals. By doing this they return vital nutrients to the soil.

GROWING BACTERIA

Bacteria multiply by cell division. One cell divides into two and then we have two bacteria instead of one. This process is called binary fission.

These two both divide and we have four. They can multiply very rapidly. Under some conditions the population can double every twenty minutes! Bacteria multiply much more slowly at lower temperatures. When food is kept in a fridge the bacteria reproduce more slowly and the food stays fresh for longer.

FUNGI

Fungi is the plural of fungus. Fungi are micro-organisms and include yeast, moulds and mushrooms. Like bacteria they play a vital role in the nitrogen cycle breaking down plant and animal remains. Also like bacteria, there are good ones and bad ones.

BAD FUNGI

Fungi cause food to go mouldy and make it unusable.

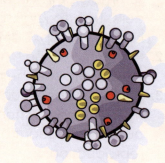

Some fungi can attack humans: athlete's foot is a fungal growth that lives between people's toes and in their toenails.

GOOD FUNGI

An obvious example of good fungi is mushrooms, which are an important foodstuff. In some cheeses the fungal growths are started deliberately to give colour and flavour.

VIRUSES

Most viruses are about a hundred times smaller than bacteria. They cannot be seen using light with a normal conventional microscope and can only be viewed with an electron microscope. Viruses can only reproduce when they are inside the cells of another organism. The common cold, flu and AIDS are among the many illnesses caused by viruses. Antibiotic drugs which kill bacteria usually do not affect viruses. You may well have heard a doctor say 'It's a viral infection and won't respond to antibiotics.'

78

ACTIVITIES

QUESTIONS

1. Name two types of micro-organism.
2. Give an example of a disease caused by bacteria.
3. Where in your body do you find good bacteria?
4. Name a food product made using bacteria.
5. How can the growth of bacteria in food be slowed?
6. Give an example of fungus that is useful to people.
7. Give an example of a fungus that is harmful.
8. How do micro-organisms contribute to the ecosystem?

AND NOW TRY

1. Protozoa cause a number of diseases, including malaria, that are common in tropical countries. Find some information about these organisms, the diseases they cause and how they are transmitted. A good place to start is http://www.biologyreference.com/Po-Re/Protozoan-Diseases.html.
2. Find out about electron microscopes and what they do. http://www.howstuffworks.com/ is a good place to start.
3. For some fun facts about bacteria look at http://ilovebacteria.com/.
4. Some bacteria have become resistant to antibiotic drugs. Find out what MRSA is and what it does.

MAKE THE LINK

English – you learn how words like bacterium and coccus form their plurals; the nursery rhyme 'Ring a Ring a Roses' has been linked to the black death.

History – you learn about the bubonic plague or black death.

RE – you read stories about people with leprosy in the Bible.

ICT – the term virus is used to describe a harmful piece of software.

Home Economics – you learn about baking and the use of yeast.

DID YOU KNOW?

The bacterium *Yersinia pestis* was probably the most deadly killer in human history. It was this that caused the plague or black death. The bacteria were transmitted to humans by fleas which were carried by rats. Another bacteria which people greatly feared was *mycobacterium leprae*, which causes leprosy. Both these diseases are now very rare because the bacteria can be killed by antibiotic drugs.

79

OUR AMAZING WORLD

Bacteria help our intestine to process waste and to absorb nutrients. They also prevent the growth of harmful bacteria. As well as helping us to digest food, some bacteria are used to prepare food. Yogurt is made by the action of bacteria on cow's milk and people have been making it for thousands of years as a way of preserving milk.

TOP TIP

Biology uses the term 'organism' to describe any living thing. A micro-organism is a very small organism that can only be seen with a microscope. Some play a vital role in the ecosystem; some are harmful to humans and some are beneficial.

DEFENCE

If you have a cold you might sneeze, cough, your nose may run and your eyes water. This is your immune system in action. It's making you do all these things to try and wash the harmful material out of your body. It might be a cold or flu virus; it might be pollen giving you hay fever, but the result is the same. Your body wants it out!

When you swallow something which contains a poison or harmful bacteria, again your body tries to flush it out. You can guess: you are sick, have diarrhoea or even both.

UNDER ATTACK!

All around us there are bacteria, viruses and other harmful organisms and substances which, if they invade our bodies, can make us ill or even cause death. These can enter our bodies in three main ways: we can breathe them in, swallow them with our food and drink, or they can enter through a cut or break in the skin. Fortunately, our bodies have an **immune system** to protect us against these invaders. This works in a number of ways.

MAKE IT HOT

When you are healthy your body temperature is 37°C, but if you are fighting an infection then your body temperature will rise by a degree or two and you become feverish. People say you 'have a temperature' or sometimes 'you are running a temperature'. It may make you feel bad, but your body is trying to make it difficult for the bacteria to live and reproduce inside you. It is one way your immune system fights infection.

WHITE BLOOD CELLS

There are two main types of blood cell: **red blood cells**, which carry oxygen round our bodies and **white blood cells**, which help us fight infection.

When organisms such as viruses or bacteria invade your body, the white blood cells get to work in two ways:

1. Some cells produce special chemicals called **antibodies** which stick to the invaders and disable them.

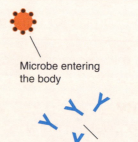

Microbe entering the body

Antibodies

2. Once the antibodies have stuck to the invading bacteria they signal to other cells called **phagocytes**. The phagocytes then come and eat the invaders.

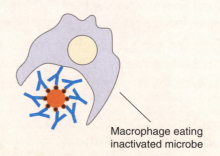

Macrophage eating inactivated microbe

TOP TIP

Harmful substances and organisms can enter our bodies by being breathed in, being swallowed, or through a cut or break in the skin. White blood cells destroy invaders by producing antibodies or by surrounding and digesting them.

80

ACTIVITIES

QUESTIONS

1. What are the three ways in which infection can enter the body?
2. What is normal body temperature?
3. What is your body trying to do when you sneeze?
4. What are the two ways in which white blood cells attack bacteria?
5. Why does your body temperature rise when you are ill?

AND NOW TRY

Use an encyclopedia or an internet search to try and find out the following:

1. What is the highest body temperature ever recorded in a human?
2. What are antibiotics and what do they do? More about this is in the famous scientists section at the end (see page 116).
3. What is the difference between bacteria and viruses?
4. Which diseases are carried by contaminated drinking water?
5. The MMR (measles, mumps and rubella) vaccination is usually given to children around their first birthday, with a booster dose just before starting school. Some parents don't want their children to have it because they are worried about the side effects of the vaccine. Look at the National Health service web site for information about MMR http://www.immunisation.nhs.uk/Vaccines/MMR.
6. Design a poster to display in a Health Centre to encourage parents to have their children vaccinated.
7. Make up a role-play activity where some people play the part of parents worried about the vaccination and the others play the part of the doctor who is trying to reassure them.

MAKE THE LINK

Geography – you learn about water-borne diseases in the developing world.

Social Education – you may discuss HIV and how to protect yourself.

Home Economics – you study nutrition and the immune system.

DID YOU KNOW?

Human Immunodeficiency Virus (HIV) is a virus that attacks the immune system. It prevents the white blood cells from doing their job properly. Infection with this virus can lead to Acquired Immune Deficiency Syndrome (AIDS), where the patient's immune system is badly weakened or destroyed. Someone who dies from AIDS is not usually killed by the HIV virus but by some other infection which their immune system is too weak to fight.

OUR AMAZING WORLD

It is quite likely that you had chickenpox when you were young and this means your white blood cells would have produced antibodies to fight the infection. Later, when you recovered, the antibodies remained in your blood. You can never catch chickenpox again because you have **natural immunity**. If you are exposed to chickenpox again your immune system will immediately recognise the virus and kill it. This happens with many illnesses: you can only catch them once and after that you are immune.

Vaccination makes use of this effect. When you are vaccinated against a disease like measles the doctor injects a dead or weakened form of the virus. This causes your body to produce antibodies so that if a real measles virus attacks you it is immediately recognised and killed by your immune system.

FERTILISATION

SPERM MEETS EGG

During sexual intercourse semen is released into the woman's vagina by the man's penis. Sperm cells in the semen swim through the uterus and into the fallopian tubes. If the sperm cells meet with an egg, cell fertilisation can take place.

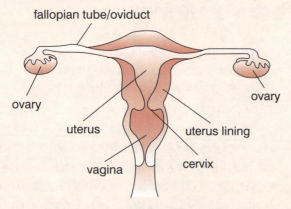

The sperm cells cluster around the egg cell.

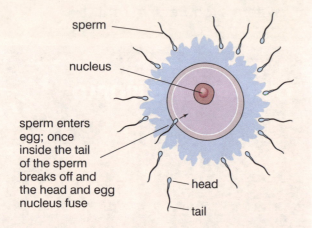

One sperm enters the egg. The egg then grows a special lining to stop other sperm entering. The sperm nucleus and the egg nucleus join together and the fertilised egg is now called a **zygote.**

The zygote divides into two cells, then four, then eight as it passes along the fallopian tube.

When this ball of cells reaches the uterus it embeds itself into the lining. At this stage it is called an embryo. The embryo continues to grow and after eight or nine weeks it is recognisably human. At this stage it is referred to as a foetus.

DEVELOPMENT IN THE UTERUS

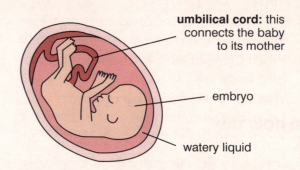

umbilical cord: this connects the baby to its mother

embryo

watery liquid

The developing embryo is connected to its mother by the umbilical cord. The baby's blood passes through the cord to the placenta. In the placenta the mother's blood and the baby's blood come close together. They do not mix and are kept apart by fine tissues.

Food and oxygen pass from the mother's blood into the baby's blood and waste products pass from the baby's blood to the mother's blood. Watery liquid called amniotic fluid surrounds the baby and protects it from bumps as the mother moves around and carries out the everyday tasks of her life. After nine months the baby is ready to be born. The walls of the uterus contract and push the baby out through the vagina.

TOP TIP

Fertilisation is when the nucleus of the sperm and the nucleus of the egg join. The fertilised egg divides to form a ball of cells and embeds itself in the lining of the uterus.

ACTIVITIES

QUESTIONS

1. What name is given to a fertilised egg?
2. What does a fertilised egg develop into?
3. What does the fertilised egg do when it reaches the uterus?
4. What protects the foetus from bumps and knocks?
5. What passes along the umbilical cord from mother to baby?
6. What passes along the umbilical cord from the baby to the mother?
7. Where do substances pass between the mother's blood and the baby's blood?
8. What harmful substances can pass into the baby's blood?

AND NOW TRY

1. After the baby is born the placenta and the umbilical cord pass out through the vagina. This is often called the afterbirth. In our society it is usually sent to the hospital incinerator. Some societies have traditional ways of dealing with the afterbirth. Do some research: what are the traditions of Maori people or native Americans?
2. Design a poster advising women not to smoke or drink alcohol in pregnancy.
3. Do an image search for pictures of babies developing in the uterus. What technology was used to get these pictures?
4. There are some very interesting stories about twins and their special relationship. Try an internet search for 'twin stories'.
5. Find out how triplets can be conceived. There are a number of ways.

MAKE THE LINK

English – you learn how to use the prefixes 'mono' and 'di' with different words.

RME – you learn about the traditions of other societies and their beliefs about childbirth.

DID YOU KNOW?

The placenta keeps the mother's blood and the baby's blood separate. Usually this means that harmful substances in the mother's blood cannot reach the baby. The mother may have a cold or flu, but the virus which causes the illness cannot reach the baby's blood. There are some exceptions:

- Rubella or German Measles is not really a serious illness for adults but it can have very serious consequences for a developing foetus. The rubella virus can pass through the placenta to the baby's blood. Children who are infected with rubella in the uterus are often born blind or deaf and with other mental and physical problems. The MMR vaccine mentioned in the section on defence protects women against rubella, so this risk has been reduced.
- Nicotine, alcohol and other drugs can also pass to the baby's blood.

83

OUR AMAZING WORLD

Twins can happen in two ways. Two eggs – one from each ovary – may be released instead of one. If both become fertilised the woman becomes pregnant with non-identical twins. They are sometimes called 'fraternal' twins or 'dizygotic' twins because they come from two zygotes.

Identical or 'monozygotic' twins come from one fertilised egg or zygote. After fertilisation the egg splits in two and each half develops into an embryo. The woman is now pregnant with identical twins.

GENETICS

CHROMOSOMES

Inside almost all human body cells there are twenty-three pairs of chromosomes, forty-six chromosomes in all. Sex cells (eggs and sperm) only have twenty-three chromosomes. At the moment of fertilisation twenty-three chromosomes from your mother's egg cell join with twenty-three from your father's sperm cell to make the forty-six needed for a new human being.

We can think of chromosomes as living libraries of information, or recipes. Along each chromosome are arranged thousands of **genes**. These control all the characteristics of a person: hair colour, height and even a tendency to suffer from some illnesses. We call this collection of recipes the *genome*. We have two copies of each gene – one from our mother and one from our father.

DNA

Each chromosome contains one very long molecule of DNA (deoxyribonucleic acid). The DNA molecule is arranged in a kind of corkscrew shape called a double helix. It stores the instructions for a fertilised egg to grow into a fully developed foetus.

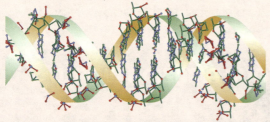

Once the baby is born the DNA contains all the information needed for it to develop into an adult: how to grow bones, hair, muscles and to bring about the changes of puberty. And finally, as well as growing our bodies, the chemical information in the DNA instructs cells in how to maintain them: how to circulate blood, digest food, heal cuts and mend broken bones.

DNA SEQUENCE

A DNA molecule is made of two long spirals (the double helix) held together by cross links a bit like the rungs on a ladder. These links are made from four types of chemical building blocks called adenosine, thymine, cytosine and guanine. Most often they are just referred to as A,T, C & G.

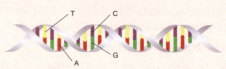

These building blocks are arranged in pairs, called base pairs, all along the molecule. The order in which these pairs are arranged carries the information stored in the DNA in the same way that the way the letters of the alphabet are arranged can spell out a word.

DNA PROFILES

Every individual's genome is unique. The sequence of the A, T, C and G building blocks in the DNA molecule is slightly different, and this sequence is called their DNA profile. Everyone has a unique DNA profile, just like they have a unique fingerprint. In fact, DNA profiles are sometimes called genetic fingerprints. They can be used in many ways.

PATERNITY OR MATERNITY TESTING

You inherit your DNA from your parents so it is possible to use profiles to identify parents, grandparents and children. DNA testing can be used to prove whether or not someone is the father of a child and is now often used to prove relationships between people and to trace ancestors. After wars or natural disasters DNA testing can be used to reunite children with their parents and grandparents and even identify bodies and body parts.

SOLVING CRIMES

In the past a criminal could be identified if he left fingerprints at the scene of a crime. DNA profiling allows a person to be identified from blood, semen, saliva, hair or dead skin. This method was first used in the mid 1980s to convict someone of rape.

84

ACTIVITIES

QUESTIONS

1. Where is DNA found?
2. What name is used to describe the shape of a DNA molecule?
3. How many pairs of chromosomes do humans have?
4. What are the building blocks which make cross links in DNA molecules?
5. What can be used to make a DNA profile of a suspect from a crime scene?
6. Where do the forty-six chromosomes come from when fertilisation occurs?
7. What is DNA profiling sometimes called?
8. Give an example of a disease that is genetically transmitted.

AND NOW TRY

1. There are some experiments to extract DNA which you can carry out in the kitchen or in the school lab. Have a look at http://www.funsci.com/texts/index_en.htm.
2. Have a look at http://learn.genetics.utah.edu/content/labs/extraction/ for some virtual experiments on DNA extraction.
3. In many countries convicted criminals have their DNA profile recorded along with their fingerprints. It has been suggested that everyone's profile be kept by the government or the police. This would make it very much easier to solve crimes. Is this desirable or does it intrude on people's privacy? Consider both points of view and organise a debate.
4. Find out about sickle cell anaemia and who is affected by it.

TOP TIP
A chromosome is a coil of DNA. Chromosomes carry the genetic instructions for all cell activities. Humans have twenty-three pairs of chromosomes.

MAKE THE LINK

English – Crime stories refer to fingerprinting and DNA profiling.

Mathematics – you learn about shapes like the helix.

RME – you can discuss and share opinions on DNA profiling.

History – DNA fingerprinting can be used to track the movements of groups of peoples in the past.

DID YOU KNOW?

Some diseases, like sickle-cell anaemia and cystic fibrosis, are hereditary. They occur when people inherit faulty genes from their parents. By examining a person's DNA it is possible to see if they have the gene or genes which cause that disease. Research is now being carried out which may make it possible to identify the damaged gene and replace it.

85

OUR AMAZING WORLD

Gender is determined by the twenty-third pair of chromosomes in your cells. Females have two X chromosomes and males an X and a Y. The defect that causes colour-blindness is carried on the X chromosome. We will call that chromosome Xc. Now let's imagine 4 people

- A male with normal vision has sex chromosomes XY

- A colour-blind male carries the defective chromosome, so his sex chromosomes are XcY. The defective Xc chromosome means that he is colour-blind.

- A female with normal vision has two chromosomes XX.

- A female carrier has one defective chromosome so her chromosomes are XcX. They have normal vision because of the second X chromosome, but can pass the defect onto their children.

SUMMARY ACTIVITIES

SMALLPOX

This project looks at one of medicines greatest success stories. Small pox was a deadly disease for centuries and has now been completely removed from the planet. The last case was in 1977.

MILKMAIDS

A milkmaid was a girl or a woman employed to milk cows. Now this work is done by machines. Things to discuss

- Why did it make sense for women to do this work?
- Does this gender stereotyping still exist?
- Can you give examples?

Milkmaids appear in many art forms. Can you find examples of where they appear in:

- A song
- Famous painting
- As the heroine in a novel
- In mythology

They were thought of as being beautiful and especially as having clear skin. There is a simile 'as smooth as a milkmaid's skin'. Can you think of any other similes used to describe people?

THE FIRST VACCINATION

So why are milkmaids such a symbol of beauty? Science has the answer. For thousands of years smallpox was one of the most serious diseases to affect humans. It killed 30% of people infected and those who survived were left with deep scars (pock marks) on their faces. But milkmaids did not seem to catch smallpox so they always had clear skin. The women who milked the cows used to be infected with a much less serious disease called cowpox. They picked up the virus from the cows while they milked them. Cowpox caused no serious symptoms and they recovered quickly and when they recovered they were then immune to the smallpox virus. It was by observing this fact that the first vaccinations were developed.

WORLD HEALTH ORGANISATION (WHO)

The world health organization now claim that small pox has been completely eradicated from the planet. The last natural case occurred in Somalia in 1977.

Some points to discuss and research:

- What is WHO and what are its aims?
- What are the main concerns of this organization today?

Here is a link to their website http://www.who.int/mediacentre/factsheets/smallpox/en.

ACTIVITIES

1. Match the key words with their meanings.

Keywords	Meanings:
a) digestion	1. An instrument for listening to sounds inside the body.
b) air sacs	2. Chemicals which stick to invading cells and disable them.
c) stethoscope	3. All the genetic information stored in a person's DNA.
d) nucleus	4. Spaces in the lungs where oxygen passes into the blood and carbon dioxide passes out.
e) micro-organism	5. The part of the cell which controls its operation.
f) unicellular	6. Consisting of one cell.
g) antibodies	7. The breaking down of large food particles into soluble substances.
h) fertilisation	8. A fertilised egg.
i) zygote	9. The joining together of an egg cell and a sperm cell.
j) genome	10. An organism which can only be seen with a microscope.

2. Class 1C2 did a survey of eye colours and got the results shown in the table below.

Eye colour	Number of pupils
Green	2
Blue	8
Brown	6
Hazel	4

Use this information to create and complete a bar chart.

Complete the following mindmap.

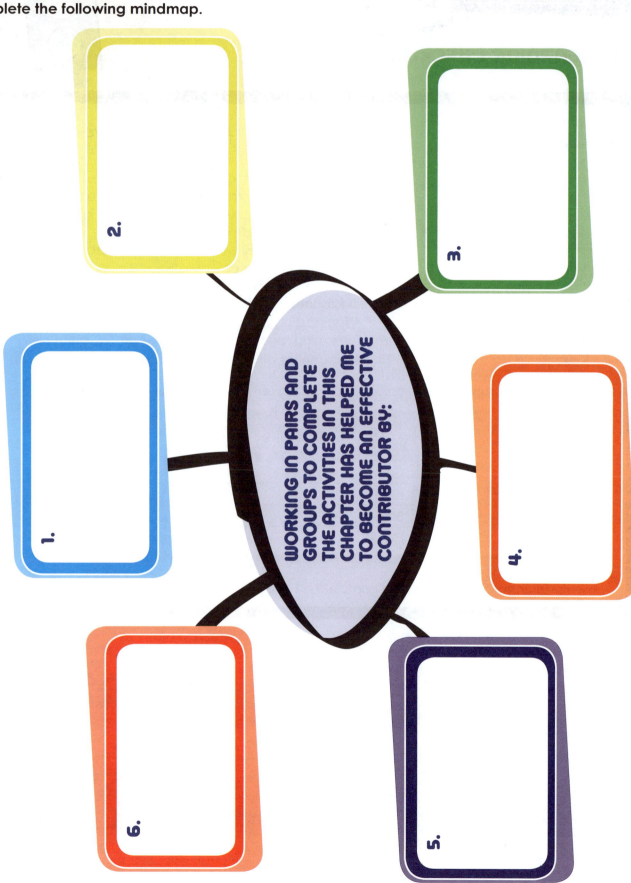

WORKING IN PAIRS AND GROUPS TO COMPLETE THE ACTIVITIES IN THIS CHAPTER HAS HELPED ME TO BECOME AN EFFECTIVE CONTRIBUTOR BY:

1.

2.

3.

4.

5.

6.

BODY SYSTEMS * BODY SYSTEMS * BODY SYSTEMS

MY PROGRESS

MY LEARNING CHECKLIST

	Help needed	Getting there	Good to go!
1. I have developed an understanding of the main systems of the human body and can explain their function and how they support life.	◯	◯	◯
2. I have explored the role of technology in the prevention, detection and treatment of disease.	◯	◯	◯
3. I have used a microscope to develop an understanding of cells and their structure.	◯	◯	◯
4. I have developed a knowledge of various types of micro-organisms and can explain how they can be both harmful and beneficial to human life.	◯	◯	◯
5. I have explored how the body defends itself against disease and can describe how vaccines can provide protection.	◯	◯	◯
6. I understand the process of fertilisation and the development of the fertilised egg into a baby.	◯	◯	◯
7. I can explain how DNA stores the instructions for building and maintaining cells and the ways in which DNA profiling can be used.	◯	◯	◯
8. I have discussed the advantages and disadvantages of DNA profiling.	◯	◯	◯

ELEMENTS

All the words in the English language are made from the twenty-six letters of the alphabet. In the same way that words are made with letters, all substances in the universe are made up from elements. There are two ways to think about this:

• An element is something that cannot be broken down into something simpler. If we take the element gold and divide it into the smallest particles, then all the particles will be gold particles.
• All substances are made up of tiny particles called atoms, and an element contains only one type of atom. Gold contains only gold atoms. Silver contains only silver atoms.

In the nineteenth century a famous Russian scientist called Mendeleev found that these elements could be organised into groups in an arrangement we now call the periodic table (see below). The columns in the periodic table are called groups and are like families of elements. Like the individuals in human families, the elements in the groups are not the same but you can see family resemblances.

THE ALKALI METALS

This is the first group in the table and starts with lithium, sodium, potassium. These are very unusual metals and react violently with water.

HALOGENS

Group 7 of the table, called the halogens, is a group of poisonous elements starting off with fluorine and chlorine. They are very reactive and form lots of compounds. Fluorine is a very poisonous gas. Compounds of fluorine, called fluorides, are found in some toothpastes.

Chlorine is also a poisonous gas. It was used as a chemical weapon in World War One and is now used in bleach and to kill bacteria in water.

Bromine is a poisonous brown liquid.

Iodine is a crystalline solid which is sometimes dissolved in alcohol to make an antiseptic.

NOBLE GASES

The noble gases in group 8 are the elements helium, argon, neon, krypton, xenon and radon. They rarely take part in chemical reactions and are sometimes called inert gases.

Helium is less dense than air and is used in balloons and airships.

Argon is used to surround the filament in a light bulb.

Neon is used to make advertising signs.

90

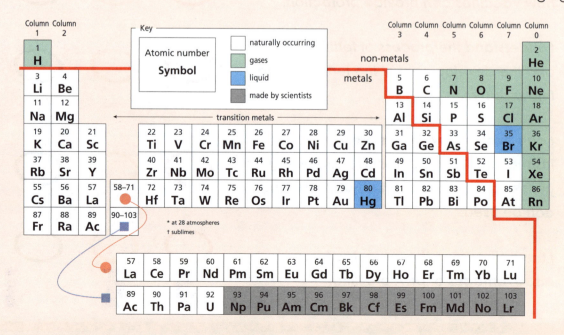

ACTIVITIES

QUESTIONS

1. What is an element?
2. Which two elements are liquids?
3. What name is given to the arrangement of elements worked out by Mendeleev?
4. Which group of elements contains the poisonous gas chlorine?
5. Which group of elements rarely takes part in chemical reactions?
6. Which group of elements reacts violently with water?
7. What division is made by the stepped line on the periodic table?

AND NOW TRY

1. In the 1950s a singer-songwriter called Tom Lehrer wrote a song containing the names of all the chemical elements. Find the 'Elements' song on You Tube. He didn't write the tune. Can you find out who did?
2. Dmitri Mendeleev lived in Russia in the nineteenth century. Search for information about him and write a short biography.
3. Elements have taken their names from lots of sources. Look for ones that are named after:
   a) famous scientists
   b) countries
   c) a state in the USA
   d) a mythological character
   e) a village in Scotland

TOP TIP
Elements cannot be broken down into anything simpler, they contain only one type of atom.

MAKE THE LINK

English – you learn to use the word 'elements' in other contexts.

Music – you may learn about the composer whose tune Tom Lehrer borrowed for the elements song.

Geography – the term 'elements' is used to describe the weather.

History – you learn about the modernisation of Russia. You learn about the use of chlorine gas in the trenches in World War One.

DID YOU KNOW?

Element facts:

- At room temperature two elements are liquids, eleven are gases and the rest are solids. They are colour-coded on the periodic table.

- Most elements are metals. Metals and non-metals are separated by the red stepped line on the periodic table.

- Ninety-two elements are naturally occurring; the rest have been made by scientists. The man-made elements are colour-coded on the periodic table.

91

OUR AMAZING WORLD

A filament light bulb has a metal filament made of the element tungsten. The space around the filament is filled with the inert gas argon so that no chemical reaction takes place.

In a halogen bulb a small amount of bromine or iodine is added and this prevents the glass from darkening due to some tungsten being left on the inside of the bulb. This means the bulb can operate at a higher temperature and halogen bulbs are smaller and more efficient than normal bulbs.

MIXTURES AND COMPOUNDS

Sulphur and iron are both elements. In the picture below sulphur powder is mixed with iron filings so the mixture now contains atoms of sulphur and atoms of iron. The two parts of the mixture are then separated using a magnet.

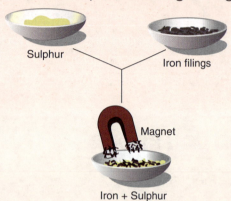

Sulphur

Iron filings

Magnet

Iron + Sulphur

The iron and sulphur have formed a **mixture** which can be easily separated.

In the next picture the iron and sulphur have been mixed in the same way but this time some heat energy has been provided to start a chemical reaction.

Heat

Iron + Sulphur

Now the atoms of sulphur combine with the atoms of iron to form molecules of the **compound** iron sulphide.

In a **chemical reaction** like this something new is formed and there is an **energy change**. Notice how the mixture glows red hot as heat energy is produced. We summarise the reaction in a word-equation like this:

iron + sulphur → iron sulphide + energy.

OTHER COMPOUNDS

When magnesium burns, the element magnesium forms a compound with the oxygen in the air. The word-equation for this reaction is:

magnesium + oxygen – magnesium oxide + energy

Again something new is formed: the compound magnesium oxide. There is also an energy change: chemical energy changes to heat and light energy.

Compounds are often completely different from the elements that form them. The elements hydrogen and oxygen combine to make water; the poisonous gas chlorine and the reactive metal sodium combine to make sodium chloride, which is common salt.

NAMING COMPOUNDS

Compounds with names ending in -ide contain only the two elements named. We have seen two so far. Iron sulph**ide** contains iron and suphur, magnesium ox**ide** contains magnesium and oxygen. We can apply this to others. Copper and oxygen form the compound called copper oxide. Copper and chlorine will form copper chloride.

Compounds with names ending in -ate contain the two elements named *and oxygen*. Copper sulph**ate** contains copper, sulphur and oxygen. Potassium chlor**ate** contains potassium, chlorine and oxygen.

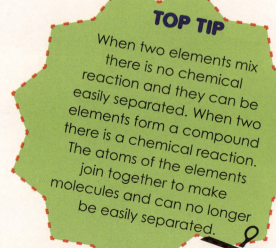

TOP TIP

When two elements mix there is no chemical reaction and they can be easily separated. When two elements form a compound there is a chemical reaction. The atoms of the elements join together to make molecules and can no longer be easily separated.

92

ACTIVITIES

QUESTIONS

1. What always happens in a chemical reaction?
2. What name is given to two or more atoms joined together?
3. What does the ending '-ide' at the end of a compound's name tell us?
4. What does the ending '-ate' at the end of a compound's name tell us?
5. Complete this table:

Element Present	Name of Compound
Magnesium and chlorine	
	Sodium fluoride
Copper, nitrogen and oxygen	
	Potassium chlorate

AND NOW TRY

1. Find a video of the iron and sulphur reaction on YouTube.

2. Rock salt, which is put on roads in the winter, is a mixture of salt and sand. Work in a group and make a plan to separate this mixture.

3. Chromatography is a technique for separating mixtures, especially coloured dyes. You can carry out simple experiments at home with coloured pens, and coloured sweets. Have a look at http://library.thinkquest.org/19037/paper_chromatography.html or do your own search.

MAKE THE LINK

Home Economics – when you are baking you start out with a mixture of the ingredients, then when you cook the mixture chemical reactions take place and you have made something new.

English – the word 'compound' can be used in other contexts: you can have a compound verb, a compound sentence or if you are very unlucky a compound fracture!

Maths – you learn about compound interest.

DID YOU KNOW?

The reaction between the elements hydrogen and oxygen is used to power the main engines of the space shuttle. The two elements react to form water:

Hydrogen + Oxygen → Water + energy

And the energy released powers the spacecraft.

93

OUR AMAZING WORLD

Air is a mixture of gases as explained in the section on gases of the air on page 26. It is mostly made up of nitrogen (about 79%) and oxygen (about 20%). Sometimes scientists want to separate these gases. Oxygen is used in hospitals and for welding apparatus, while nitrogen can be used for making fertilisers and explosives. If air is made very cold, all the gases turn into liquids. The liquid is slowly allowed to warm up and at −196°C the nitrogen will turn back to a gas and can be collected separately. The oxygen does not turn back to a gas until −183°C so the two main gases have now been separated.

SEPARATING COMPOUNDS

We saw in the last section that when two elements combine to form a compound there is an energy change. The atoms of the two elements join and cannot be easily separated. The atoms bond together to make molecules. To separate the elements in a compound energy must be supplied to break the bonds.

HEAT

Some compounds will separate into their elements when heated. Mercuric oxide is a compound of mercury and oxygen. When heated it breaks apart into the elements mercury and oxygen:

mercuric oxide + heat → mercury + oxygen

Mercuric oxide is both expensive and highly poisonous, so not many people have tried this. However, it was by carrying out experiments like this that the French scientist Antoine-Laurent de Lavoisier discovered the existence of oxygen and explained the process of burning.

ELECTRICITY

Many compounds can be separated with electrical energy. This process is called **electrolysis**. Water is a compound of hydrogen and oxygen and in the experiment below an electrical current from the battery passes through the water and bubbles of gas appear as the compound is split into hydrogen and oxygen. You will notice in the diagram that

there is roughly twice as much hydrogen as oxygen. This tells us that the water molecules contain two atoms of hydrogen and one of oxygen and explains the chemical name for water: H_2O.

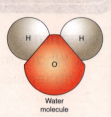
Water molecule

Other compounds can be separated by electrolysis. Copper chloride is a green crystalline solid which can be dissolved in water. See pages 96 for information on dissolving. The process of electrolysis can be used to split the compound into the two elements copper and chlorine.
The reaction can be summarised in a word-equation:

copper chloride + energy → copper + chlorine

Again, notice that the compound is quite different from the elements that combine to form it. Copper is a reddish-coloured metal and chlorine is a gas.

LIGHT

Sometimes light energy can break the bonds in a molecule and separate a compound into its elements. Light-sensitive film for cameras uses a compound of the metal silver and bromine, an element of the halogen family. When light strikes the compound the compound breaks up into its elements:

silver bromide + light → silver + bromine

The metal silver gives the dark image on the film.

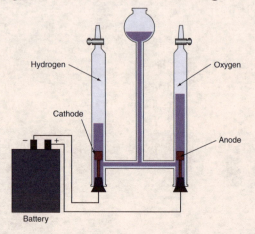

Hydrogen

Oxygen

Cathode

Anode

Battery

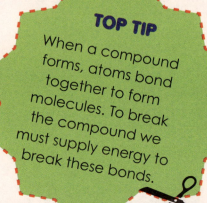
TOP TIP
When a compound forms, atoms bond together to form molecules. To break the compound we must supply energy to break these bonds.

94

ACTIVITIES

QUESTIONS

1. What form of energy did Lavoisier use to separate the compound mercuric oxide?

2. What must always be supplied to separate the elements in a compound?

3. What causes the molecules in plastics to break apart?

4. What causes the darkening of photographic film when light shines on it?

5. Compounds are very different from the elements that combine to make them.
   a) Which elements combine to make copper chloride?
   b) Describe copper chloride.
   c) Describe the elements that make up copper chloride.

6. What form of energy separates water into hydrogen and oxygen?

7. Which compounds can be separated by light?

AND NOW TRY

1. Some paints are damaged by ultraviolet rays. Have a look for UV degradation of car paint. Which colour seems to be affected most?

2. The stratosphere is a layer in the earth's atmosphere:
   a) how high is the stratosphere?
   b) what are the names of the other layers of the atmosphere?

3. Draw a diagram of the layers of the atmosphere

4. Some human activity is damaging to the ozone layer. Find out which man-made chemicals are doing the damage and design a poster to raise awareness.

MAKE THE LINK

History – Antoine-Laurent de Lavoisier was executed by guillotine in 1794 during the French Revolution. You may learn about the French Revolution.

CDT – you learn about the strength and use of materials. Materials which are exposed to sun must be able to withstand ultraviolet radiation.

Art – you learn about colour and the effects of light.

Geography – you learn about the structure of the earth and its atmosphere.

DID YOU KNOW?

The effect of light (especially ultraviolet light) on compounds can be damaging. Plastics are made of very long molecules called **polymers** and if they are exposed to sunlight the energy of the ultraviolet rays breaks the bonds in the compound. This weakens and discolours the material. This process is called **UV degradation**.

OUR AMAZING WORLD

In the air, atoms of oxygen combine in pairs to make molecules. Ultraviolet radiation from the sun can break the bond in the molecule and give two separate atoms.

The single atoms of oxygen then go on to form molecules of ozone which contain three oxygen atoms.

The ozone absorbs the most of the ultraviolet rays and stops them reaching the earth's surface.

DISSOLVING

SOLUTIONS

When a solid dissolves in a liquid it seems to disappear into the liquid and the final liquid is clear. It may be coloured, but it is clear – or transparent – so we can see through it.

There are some important words to get right here.
The solid we put into the liquid is called the **solute**.
The liquid we're putting the solid in is called the **solvent**.
The result (the solute dissolved in the solvent) is called the **solution**.

For example, when we dissolve the copper sulphate in water the blue liquid we have at the end is called a **solution**. The liquid – the water – is called the **solvent**. The solid which

solute + solvent → solution

dissolves is called the **solute**.

solute + solvent → solution

There are three ways to speed up the process of dissolving: heat the water, use finer particles of solid and stir the liquid.

SATURATED SOLUTIONS

If we keep adding more solute to a solution we will come to a point where it is just not possible to dissolve any more. We have now produced a **saturated solution**. It simply will not dissolve any more solid. We can try adding more, but it will just fall to the bottom.

EFFECT OF TEMPERATURE

Most solids become more soluble as the temperature increases. This means that a given volume of solvent will be able to dissolve more solute. Suppose you have a saturated solution of copper sulphate at room temperature. If you heat it you will then be able to dissolve more solid.

It also works the other way around. If you cool the saturates solution crystals will form, because less solid can dissolve at the lower temperature.

Cooling more slowly produces larger crystals.

TOP TIP

Solids which dissolve are said to be soluble. Things which will not dissolve are insoluble. Solubility increases when the temperature increases.

96

MATERIALS * MATERIALS

ACTIVITIES

QUESTIONS

1. When you boil pasta you add salt to the water. In this case what are the solute, the solvent and the solution?
2. What are three ways in which you can make a solid dissolve faster in a liquid?
3. Alan is carrying out an experiment to find how temperature affects dissolving. What two things must he keep constant to make his investigation fair?
4. What size of crystals would you expect if you were to cool a saturated solution quickly?
5. What name is given to a solution which will not dissolve any more solute?

AND NOW TRY

1. You can buy a crystal-growing kit quite cheaply on eBay or at Amazon. Try growing your own crystals.
2. Fill a kettle with cold water and listen as it boils. Some time before it boils you will notice some sounds coming from it. People sometimes talk about the kettle 'singing'. This happens because the dissolved gases are being driven out of the water.
3. Let the kettle cool and boil it again. This time it doesn't sing. Can you explain why?
4. Let the water in the kettle cool completely and taste it. Not very nice! The dissolved gases have all gone and the water doesn't taste the same.
5. Precious stones like diamonds are crystals which formed as the earth cooled. You won't find them walking on the beach but you might find other crystals in stones. You might see quartz, garnet or mica crystals, depending on where you live.

MAKE THE LINK

CDT – you learn about solvents in paint and varnish.

Social Education – you learn about solvent abuse.

Geography – you learn about rocks cooling and the formation of the earth.

DID YOU KNOW?

Some things don't dissolve in water but they dissolve in another liquid. These liquids are sometimes called **non-aqueous solvents**. Non-aqueous just means 'not water'. A good example is nail varnish. It doesn't dissolve in the shower or when you wash your hands but it does dissolve in nail varnish remover because this contains a non-aqueous solvent called acetone.

OUR AMAZING WORLD

Gases also dissolve in water. Unlike solids, raising the temperature lowers the solubility, so if you heat the water the gas comes out of the solution. This explains a number of things.

- Fizzy drinks like cola contain dissolved carbon dioxide. They are cold when you drink them then warm up in your stomach. The gas comes out of solution and you burp!

- Fish breathe the oxygen which is dissolved in the water just as we breathe the oxygen from the air. The pumps which are used in fish tanks help to dissolve more oxygen in the water. This happens naturally in fast-flowing rivers. Water which is polluted and isn't flowing contains little dissolved oxygen and this makes it more difficult for fish and other organisms to survive.

THE EARTH

The earth is thought to be made up of layers as shown in the diagram:

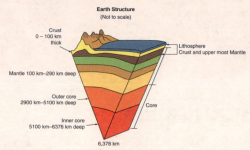

Earth Structure
(Not to scale)

Crust
0 – 100 km thick

Lithosphere
Crust and upper most Mantle

Mantle 100 km–290 km deep

Outer core
2900 km–5100 km deep

Core

Inner core
5100 km–6378 km deep

6,378 km

The core is mostly made of iron and nickel with some traces of other elements. The inner core is solid and the outer core is liquid. The liquid metal outer core spins around the inner core and this causes the earth's magnetic field.

The mantle is the thickest layer in the earth and is made of soft rock called magma. Because of the very high temperature and pressure the magma behaves like a liquid, as we've seen in the section on energy transfer and buildings (page 14), hot liquids create convection currents. Convection currents in the magma affect what happens in the layers above.

The uppermost layer, **the crust**, is where we live. Here we find all the materials we need for our lives. The crust is between 5 and 70 km thick. It is not one solid piece but is made up over several large pieces called tectonic plates which float on the mantle underneath.

Earth's crust facts:
• movement of the earth's tectonic plates is what causes earthquakes
• where magma breaks through an area of the earth's crust we have volcanoes.

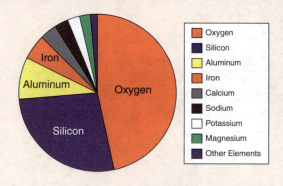

Pie chart legend:
- Oxygen
- Silicon
- Aluminum
- Iron
- Calcium
- Sodium
- Potassium
- Magnesium
- Other Elements

Almost all the ninety-two naturally occurring elements are found in the earth's crust, but mostly in tiny amounts. The main ones are shown in the chart below.

TYPES OF ROCK

There are three main types of rock. The diagram below shows how they were formed.

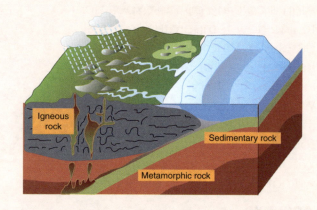

Igneous rock

Sedimentary rock

Metamorphic rock

IGNEOUS ROCK

Igneous rocks were formed from the molten magma as it bubbled out onto the earth's surface. As it cooled, crystals were formed and these can often be seen in the rocks. Igneous rocks such as granite are usually very hard. Diamonds are crystals of carbon formed as the rock cooled.

SEDIMENTARY ROCK

Sedimentary rock was formed from grains of sand that were washed into the sea or into riverbeds. Over time, more and more layers of these fine grains collected on top of each other. The weight of the layers above cemented the grains together to form rocks like sandstone. Oil and gas deposits are found in sedimentary rocks like sandstone.

METAMORPHIC ROCK

Metamorphic rock was originally igneous or sedimentary rock that has been metamorphosed (changed) due to heat or pressure. It often has a squashed appearance like this slate.

MATERIALS * MATERIALS

ACTIVITIES

QUESTIONS

1. What is the core of the earth made from?
2. Where would you find magma?
3. How is a volcano formed?
4. Which are the two most plentiful elements in the earth's crust?
5. What causes earthquakes?
6. In which type of rock do you find crystals?
7. Which type of rock has been changed by heat or pressure?
8. Which type of rock is sandstone?

AND NOW TRY

1. Search for images of the aurora borealis. What are some of the myths and beliefs about it? Have a look at http://www.luminarium.org/mythology/revontulet.htm.
2. Where in Scotland could you travel to see the northern lights? Plan a trip.
3. Geology – the study of the structure of the earth is a branch of science in its own right. Depending on where you live there will be interesting things to see. Have a look at http://www.scottishgeology.com/
4. Find out which places in the world have the most earthquakes. Mark them on the map of tectonic plates. What do you notice?

TOP TIP

The three sections of the earth are called the core, the mantle and the crust. When magma from the mantle bubbles through the crust we have a volcano.

MAKE THE LINK

Geography – you learn about the forces that shape the landscape.

English – you learn about myths and legends. Some people used to think that the earth was hollow and that there were other lands inside the hollow earth. They wrote stories about it such as Jules Verne's *Journey to the Centre of the Earth*.

History – articles and people buried in volcanic eruptions give valuable insight into life in the past.

Language – the word 'metamorphic' comes from the word 'metamorphosis', meaning a change. You learn to use words like this.

DID YOU KNOW?

California in the western USA is well known for earthquakes. This is because it is on the boundary between two tectonic plates. Earthquakes tend to happen on the edges of the plates. Jerky movements at the plate edges cause earthquakes.

99

OUR AMAZING WORLD

One of the most spectacular light shows on earth is the northern lights or aurora borealis. It is caused by charged particles from space being trapped by the earth's magnetic field.

EARTH'S OCEANS

Oceans cover 70% of the earth's surface and it is here that life first appeared on the planet. Living things did not emerge onto the land until much later. The earth's oceans play an important part in keeping the temperature of the planet stable; they are a source of food and minerals and produce at least half of the earth's oxygen.

PLANKTON

Plankton are micro-organisms that drift in the oceans and are the source of energy for most marine life. We saw in the section on photosynthesis on page 10 how green plants capture the sun's energy and use it to make food. We summarised the situation in a word-equation:

water + carbon dioxide + light ← glucose + oxygen

100

In fact only about half of the earth's oxygen is produced by photosynthesis in green plants on land. The other half is produced by photosynthesis in microscopic organisms called **phytoplankton**. Just as food chains on land start with green plants, food chains in the oceans start with phytoplankton.

Zooplankton are slightly larger creatures that feed on other plankton, so we have a food chain with phytoplankton at the start.

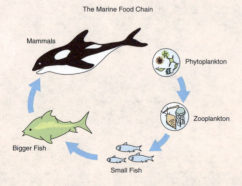

The Marine Food Chain

Mammals · Phytoplankton · Zooplankton · Small Fish · Bigger Fish

Almost everything that lives in the sea depends on the sun's energy which is captured by the tiny phytoplankton floating on the surface of the water.

MINERALS FROM THE SEA

We think of the sea as salty, but sea water in fact contains many compounds and over fifty different elements have been identified in sea water. Sea water is salty because salt and other minerals are continuously being washed off the land into the oceans and seas. The water evaporates and returns to the land as rain. The salt and minerals cannot leave, so the sea is gradually becoming more and more salty.

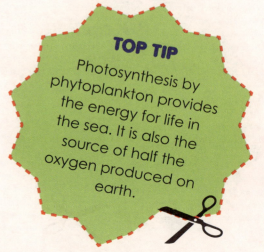

The Water Cycle
(The Hydrologic Cycle)

Condensation · Precipitation · Condensation (Clouds form)
Transpiration · Subsurface (underground) Runoff · Surface Runoff · Evaporation · Accumulation

People have obtained salt by evaporating sea water for thousands of years and you may well have sea salt in your kitchen at home. Apart from magnesium most of the other minerals are only present in very small quantities and it is not worth extracting them. About half of the magnesium we use is extracted from sea water by the process of electrolysis.

As well as dissolved salt and minerals rivers also wash sand gravel and other materials into the seas and dump them close to the coast. These are called 'placer deposits' and they are a source of many minerals including gold, tin, titanium and diamonds.

TOP TIP

Photosynthesis by phytoplankton provides the energy for life in the sea. It is also the source of half the oxygen produced on earth.

MATERIALS * MATERIALS

ACTIVITIES

QUESTIONS

1. What are the two main sources of the earth's oxygen?
2. Which type of plankton carries out photosynthesis?
3. Which type of plankton feeds on other plankton?
4. What percentage of the earth's surface is covered by water?
5. Which metal is extracted from sea water by electrolysis?
6. What name is given to minerals deposited in the sea by rivers?
7. How can energy be obtained from the sea?
8. Name two minerals mined from coastal deposits.

AND NOW TRY

1. Glow sticks produce light by a chemical reaction. You can buy them quite cheaply. Check out http://science. howstuffworks.com/light-stick.htm for an explanation of how they work.
2. Bioluminescence is mostly found in sea creatures. See if you can find any information about it in other animals.
3. Look at a packet of sea salt to see which compounds it contains.
4. The BBC series *Blue Planet* is a series of eight programmes about the oceans. You can find info at http://www.bbc.co.uk/nature/ programmes/tv/blueplanet/
5. Most scientists agree that increased carbon dioxide in the atmosphere is a cause of climate change. It has been suggested that adding chemicals like iron to the sea would increase the number of phytoplankton and so reduce the carbon dioxide. Do some research and find out what is being suggested. Discuss and develop a point of view.

MATERIALS

MAKE THE LINK

Business Education – fish farming is an important part of Scotland's economy.

English – you may read the famous poem about the salty sea:

'Water, water, everywhere,
And all the boards did shrink
Water, water, everywhere,
Nor any drop to drink.'

Geography – you learn about the water cycle.

Modern Studies – you learn about oil exploration and energy sources.

DID YOU KNOW?

In the section on energy we saw that scientists and engineers are working on ways to extract energy from the waves and the tidal movement of our seas. The fossil fuels oil and gas were originally found on land-based sites but in the 1970s we started to extract them from the North Sea. Now we are looking for ways to obtain these fuels from deeper and more remote areas of the oceans.

OUR AMAZING WORLD

Bioluminescence is light produced by a chemical reaction. A chemical called luciferin combines with oxygen and light is emitted. Like other chemical reactions it can be summarised in a word-equation.

luciferin + oxygen → oxyluciferin + light

It is thought that about 90% of the creatures that live in the deep sea create light in this way.

METALS AND ORES

METALS FROM THE EARTH

There are ninety elements found in nature, and over two thirds of these are metals. There are some very reactive metals like sodium which takes part in lots of chemical reactions and will react violently with water. Other metals like gold are very unreactive and rarely form compounds. Only very unreactive elements like gold and silver are naturally found in their pure state. All the other metals are found as **ores**. An ore is the compound of a metal with other elements, often oxygen.

Some common ores are:

Name of ore	Metal	Compound of
haematite	iron	iron and oxygen
bauxite	aluminium	aluminium, oxygen and hydrogen
chalcopyrite	copper	copper, iron and sulphur
ilmenite	titanium	titanium, iron and oxygen

EXTRACTING METALS FROM ORES

Before the metal can be used it must be **extracted** from its ore. One process for doing this is called **smelting**. This can often be done by mixing the ore with carbon then heating the mixture in a large container called a blast furnace. For example, when iron ore is heated with carbon the following reaction takes place:

iron oxide + carbon → iron + carbon dioxide

The carbon displaces the iron from the compound and we are left with the metal iron, which is used to make steel.

EXTRACTION OF METALS BY ELECTOLYSIS

This method of extraction does not work for all metals. We saw on page 94 how the process of electrolysis can be used to separate the elements in a compound. This is done on an industrial scale to extract more reactive metals like aluminium and magnesium from their ores. The picture below shows how it is done.

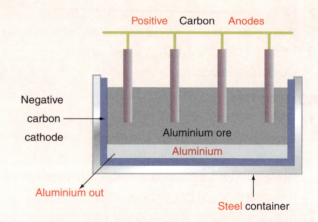

This process uses vast amounts of electrical energy and the aluminium smelters often have their own hydroelectric power stations.

TOP TIP

Remember, most metals are found as ores. An ore is a compound of the metal with other elements. Metals are extracted from their ores by heating with carbon or by using electricity.

102

MATERIALS * MATERIALS

ACTIVITIES

QUESTIONS

1. What is an ore?
2. Which metals are found uncombined in the earth's crust?
3. What metal is extracted from haematite?
4. What is heated in a blast furnace?
5. Write a word-equation for the smelting of iron in a blast furnace.
6. What ore is used as a source of titanium?
7. What name is given to the process of separating a compound by electricity?
8. Why can aluminium not be extracted from its ore by heating with carbon?

AND NOW TRY

1. Plan a day out gold panning – try a web search for gold prospecting in Scotland.
2. We often hear about the rainforests and how they are destroyed to create farmland. Not so many people realise that the forests are also cleared to mine for minerals and ore. Search for some information on this and design a poster to raise awareness of the problem.
3. The mineral malachite is an attractive semi-precious stone. It is in fact a compound of copper. Search for some more images and design a piece of jewellery or an ornament to be made from malachite.
4. Search for images of Loch Laggan and Loch Laggan dam. Why was the dam built and why is the water needed?

MAKE THE LINK

CDT – you learn about materials and tools.

Geography – you learn about extracting the earth's resources.

Art – you learn about designing jewellery.

History – you learn about the Gold Rush, when people flocked to the western United States to pan for gold in the rivers.

DID YOU KNOW?

Miners used to find gold particles in river beds by a technique called 'panning'.

OUR AMAZING WORLD

103

The element titanium was discovered in 1791. Titanium is as strong as steel but very much lighter and in the 1950s and 1960s the Soviet Union started to use it to make submarines and high-performance jet aircraft. Most of the titanium produced today is used in the aircraft industry and the four engines of the airbus contain about 26 tonnes of the metal.

Titanium has now become part of our world and examples of its use are all around. It is used: in technology to make cutting tools; in medicine to make artificial joints and dental implants; in leisure for jewellery and sports equipment like bicycle frames.

Architects have found ways to use the material in buildings and the museum in Bilbao, northern Spain, is covered in titanium panels.

CHEMICAL AND PHYSICAL CHANGES

PHYSICAL CHANGES

We saw in the section on heating and cooling matter (see page 22) that nearly all substances expand when they are heated and contract when they are cooled. This is an example of a **physical change.** The change immediately reverses when we cool the substance down again and we have not made a new substance. Other examples of physical changes are melting and evaporation. Water may evaporate into steam but it easily changes back when we cool it down.

CHEMICAL CHANGES

In a **chemical change** a new substance is formed and the change is not easily reversed. In the mixtures and compounds section (page 92) we saw how iron and sulphur react to form iron sulphide. This is a chemical change or a chemical reaction. The atoms of sulphur and iron have joined together to form molecules; a new substance has been formed and energy would be needed to separate the elements in the compound.

SIGNS OF A CHEMICAL REACTION

In a chemical reaction something new is always formed and there is always an energy change. Some other things can happen that tell us there is a chemical reaction taking place:

- there may be a colour change
- a solid may form and sink to the bottom of the container. We call this precipitation.
- bubbles of gas may form. We call this effervescence.

SPEEDING UP REACTIONS

In the picture below a chemical reaction is taking place between the chalk and the acid. Bubbles of gas are being formed quite slowly. If we wanted to increase the speed of the reaction so that the bubbles were released

faster there are three ways in which we could do this:

bubbles of gas produced slowly

bubbles of gas produced more quickly

hydrochloric acid

small lumps of chalk

small lumps of chalk

- Use smaller particles of chalk. Grind it up so that we have powder rather than a lump.
- Increase the temperature of the chemicals. Warm up the acid.
- Use a stronger acid.

CATALYSTS

Catalysts are substances that speed up chemical reactions. When you add a catalyst the chemicals react faster but the catalyst itself is not used up.

Car exhausts contain a number of harmful gases, and catalytic converters help to convert these back into harmless gases: oxides of nitrogen are turned back into oxygen and nitrogen, and poisonous carbon monoxide is turned into carbon dioxide.

Our bodies produce biological catalysts called enzymes which aid the chemical reactions involved in digestion.

TOP TIP
Some chemical reactions take place slowly and some are fast. The speed of a reaction can be increased by increasing temperature, decreasing the particle size (powder instead of lumps), increasing concentration (using stronger acid) and using catalysts.

104

MATERIALS * MATERIALS

ACTIVITIES

QUESTIONS

1. All chemical changes involve a new substance being formed and an energy change. What are the other three signs of a chemical reaction?
2. Give three ways to speed up a chemical reaction.
3. Arrange these chemical reactions in order of speed with the slowest first: rusting; magnesium burning; a cake baking; digesting food; and chalk bubbling in acid.
4. What do catalysts do?
5. What name is given to a biological catalyst?
6. What happens to the atoms of iron and sulphur when iron sulphide is formed?
7. Divide the following into physical and chemical changes: water evaporating; gas burning; ice melting; sugar dissolving; iron rusting; and an egg frying.

AND NOW TRY

1. Search for a draft of Alfred Nobel's will.
2. Find out about the Nobel prizes.
   a) How many are there?
   b) How often are they awarded?
   c) Who were the most recent people to receive them?
3. The metal used to make a statue has reacted with substances from the air and sea water.
   a) Find out what name is given to this colouring.
   b) What compound has been formed?
   c) Can you find other examples near your home?
4. We don't normally think of flour as an explosive, but because it is made of very fine particles it can sometimes cause explosions. Search for videos of flour bombs.

MAKE THE LINK

English – you learn to use words like effervescence. You might in a story describe a person as being bubbly. You could describe them as effervescent.

Modern Studies – you learn about winners of the Nobel Peace Prize.

Geography – precipitation is used to describe falling rain or snow.

CDT – you learn how materials are protected from the weather.

DID YOU KNOW?

Chemical reactions don't just take place in science laboratories, they affect all of our lives. When we digest our food chemical changes take place in the stomach and in the intestines. When we make a meal chemical changes take place in the food as it is cooked. Outside, chemical reactions take place in materials due to the weather, for example, iron and steel rust.

When iron and steel rust a chemical reaction takes place between the iron and oxygen from the air. It can be summarised by the word-equation,

iron + oxygen $\rightarrow$ iron oxide

105

OUR AMAZING WORLD

An explosion is a chemical reaction that takes place very quickly. Heat, energy and gas are released in a very short time – often less than one second. Dynamite was invented in 1867 by the Swedish engineer Alfred Nobel. It is made by soaking an absorbent material like clay with the explosive nitroglycerine.

Nobel was accused of becoming very rich by finding ways to kill people and when he died he left almost all his wealth ($200 million in today's money) to fund the Nobel Prizes.

ACIDS AND ALKALIS

THE PH SCALE

Acids and **alkalis** are very reactive chemicals. How strong or reactive these chemicals are is measured on the pH scale. This scale runs from 1 to 14, and all substances can be placed somewhere on it. Here are some examples:

Substance	pH
strong acid (from science lab)	1
strong alkali (caustic soda – used for clearing blocked drains)	14
weak acid (vinegar)	4
weak alkali (dishwasher powder)	9
neutral (water or salt water)	7

Neutral substances like water are unreactive and have a pH of 7.

ACID FACTS

- Acids have a pH less than 7.
- Acids are corrosive, which means they eat away or dissolve other substances. People talk about acid burns.
- Some common laboratory acids are:
  - hydrochloric acid
  - sulphuric acid
  - nitric acid
- When acids are present in foods they tend to taste sour, for example lemon juice, vinegar.

ALKALI FACTS

- Alkalis have a pH higher than 7.
- Alkalis are also corrosive.
- Some common laboratory alkalis are:
  - sodium hydroxide
  - potassium hydroxide
  - ammonia
- Alkalis in the home are often used in cleaning products. Oven cleaner and dishwasher powder are both strong alkalis.
- Alkalis are also called bases.

MEASURING pH

pH can be measured with **universal indicator**, which can be either a liquid or a strip of paper.

The chemical to be tested is added to the indicator and by matching the colour of the indicator with a chart like the one below it is possible to tell the pH of the liquid.

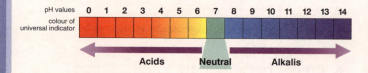

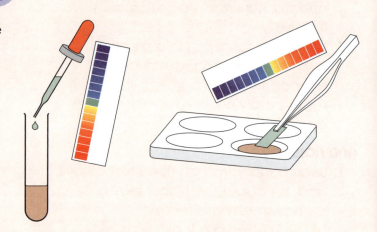

NEUTRALISATION

Neutralisation is when an acid and an alkali cancel each other out. Adding alkali to an acid makes the acid less strong. It increases the pH. Adding the correct amount will produce a neutral liquid with pH 7. Mixing hydrochloric acid and the alkali sodium hydroxide produces sodium chloride (common salt) and water.

hydrochloric acid + sodium hydroxide →
sodium chloride + water

106

MATERIALS * MATERIALS

ACTIVITIES

QUESTIONS

1. How would you describe a substance with pH 4?
2. What name is given to the cancelling out of an acid with an alkali?
3. What would be the pH of a strong alkali?
4. What is the name of the substance which is used to test for pH?
5. What colour on the pH chart corresponds to a strong acid?
6. If the indicator turns blue what does it tell us about the substance?
7. What colour would the indicator give for a neutral substance?
8. An acid is described as 2M. What information does this give you about the acid?

AND NOW TRY

1. If you boil red cabbage the cooking water which you normally pour away acts as an indicator. Try adding some common household substances like vinegar, washing-up liquid, baking soda and others to the cooking liquid.
2. Look for information on pH and shampoo. Why does baby shampoo not sting your eyes?
3. Many foods and drinks contain acids. Look at the contents list on some food and drink containers and make a list of the acids that you find.
4. Look at a packet of antacid tablets. What chemical do they contain?

TOP TIP

pH 7 means the substance is neutral. pH higher than 7 and it is an alkali. As the pH increases the strength of the alkali increases. pH lower than 7 tells us it is an acid. As the pH decreases the strength of the acid increases.

MATERIALS

MAKE THE LINK

Geography – you learn about acid rain and its effect on the environment.

Home Economics – you learn about different substances in the home.

English – the word 'corrosive' can be used in everyday speech to describe something or someone as unpleasantly sarcastic.

DID YOU KNOW?

Molarity is a measure of the concentration of an acid or an alkali. In your science lab at school you may do experiments with 1 Molar (1M) acid. 2M acid is twice as strong and may be considered safe for pupils to use. Your teacher may do a demonstration with 16M acid but would put it safely away in a chemical store as soon as they had finished.

Adding water to acid or alkali decreases its strength and it will eventually become neutral.

107

OUR AMAZING WORLD

Many people believe that if you are stung by a wasp the poison it injects into you is an alkali. If you don't have special cream for treating it you can rub on vinegar to neutralise the alkali and make it less painful. Is it true or is it just folklore? Check out http://www.insectstings.co.uk/sting-acid-or-alkali.shtml.

Indigestion is often caused by excess stomach acid. Taking antacid tablets neutralises the acid and reduces the pain.

MORE ABOUT ACIDS

ACIDS AND METALS

Many metals react with acid to produce hydrogen gas.

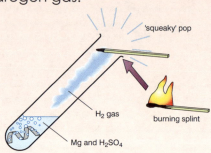

'squeaky' pop

H_2 gas

burning splint

Mg and H_2SO_4

If a burning splint is brought up to the top of the test tube the hydrogen gas burns with a squeaky pop. This test is used to identify hydrogen.

The equation can be summarised as follows:

hydrochloric acid + magnesium → magnesium chloride + hydrogen

ACIDS AND CARBONATES

Carbonates are compounds of a metal with carbon and oxygen. Calcium carbonate is found in the earth as chalk or limestone. Copper carbonate, commonly known as malachite, is a source of copper (see section on metals and ores page 102).

When carbonates react with acids they produce the gas carbon dioxide.

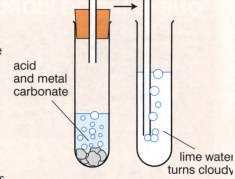

CO_2 gas

acid and metal carbonate

lime water turns cloudy

Geologists can use this test to identify rocks.

For example placing a drop of acid on a limestone pebble would cause a reaction and release bubbles of carbon dioxide. However if a drop of acid was placed on a quartz pebble, the quartz will give no reaction.

PH AND PLANTS

The soil in a garden or field is rarely neutral with a pH of 7. It is often either slightly acidic or alkaline and this affects how nutrients are absorbed by the roots. Nitrogen is best absorbed when the pH of the soil is 5·5, while phosphorus is absorbed best from a soil with a pH between 6 and 7 (see the section on chemicals and agriculture page 12 for more information on the chemical nutrients that plants need). Some plants, like rhododendrons, grow best when the soil is slightly acidic and others, such as chrysanthemums, prefer alkaline soil.

Gardeners and farmers control the pH of soil by adding lime to make it more alkaline or ammonium sulphate to make it more acidic.

The colour of some flowers is affected by the pH of the soil. Hydrangeas are quite common; lots of people have them in their gardens. They are deep blue if the soil is acid but are pink when the soil is alkaline. Like some other flowers, they are a naturally occurring indicator. Some gardeners adjust the pH of the soil in their gardens to produce attractive flowers.

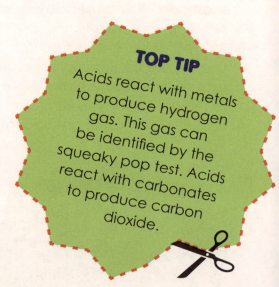

TOP TIP

Acids react with metals to produce hydrogen gas. This gas can be identified by the squeaky pop test. Acids react with carbonates to produce carbon dioxide.

ACTIVITIES

QUESTIONS

1. What is produced when a metal reacts with an acid?
2. Write a word-equation for the reaction of zinc with hydrochloric acid.
3. What is produced when a carbonate reacts with an acid?
4. What elements are in the compound calcium carbonate?
5. What elements are in the compound magnesium chloride?
6. How can a geologist distinguish between quartz and limestone?
7. How do gardeners increase the pH of soil?
8. Which nutrient is absorbed when the soil pH is between 6 and 7?

AND NOW TRY

1. You can buy a kit to test soil pH quite cheaply at a garden centre. If you have a garden or even pot plants try testing the soil.
2. You can do some interesting home experiments with vinegar and baking soda. Have a look at http://www.exploratorium.edu/science_explorer/bubblebomb.html.
3. Do some research on the internet to find plants that grow well in acid soil and ones that grow well in alkaline soil.
4. Find out which lakes are naturally acidic and why.
5. Find out more about the plants which grow well in acid soil and those which grow in alkaline soil.

MAKE THE LINK

Geography – you learn about different types of rocks; you also learn about acid rain and other forms of pollution.

RME – you discuss issues like the death penalty (see Did you know? below).

DID YOU KNOW?

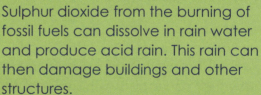

Sulphur dioxide from the burning of fossil fuels can dissolve in rain water and produce acid rain. This rain can then damage buildings and other structures.

The acid rain then runs into rivers and lochs making them slightly acid. This is clearly harmful to the plants and animals, and sometimes the water can become so acidic that nothing can live in it. The sulphur dioxide produced in the UK doesn't just affect us: some of the rain blows across the North Sea.

109

OUR AMAZING WORLD

James George Haigh, known as the acid bath murderer, was convicted of the murder of six people and executed in 1949. He actually claimed to have killed nine people. Thinking, wrongly, that the police could not bring a charge of murder if there was no body, he would place his victim's body in a large drum and then fill it with concentrated sulphuric acid. A few days later there was nothing left but sludge, and he would pour this down a drain. One of the victims was identified because his false teeth were not dissolved by the acid.

REACTIVITY

The group in the periodic table called the alkali metals contains some very unusual elements. Sodium reacts violently with water to produce the alkali sodium hydroxide and hydrogen gas is released. This can be summarised in the word-equation,

sodium + water → sodium hydroxide + hydrogen

The reaction releases heat energy and this can cause the hydrogen to ignite. The results can be quite spectacular.

Potassium reacts even more violently with water. We say it is more reactive than sodium. In the section on speed of reactions we saw that iron reacts with water but the reaction is very slow and happens over several months. Metals like gold and silver do not react at all with water. By observing the reactions of metals with water, oxygen and dilute acids it is possible to make a league table or reactivity series with the most reactive metals at the top and the least reactive at the bottom. Scientists call this the reactivity series and it goes like this:

Most reactive

potassium
sodium
calcium
magnesium
aluminium
zinc
iron
lead
tin
copper
silver
gold

Least reactive

METALS AND OXYGEN

Some metals react with oxygen from the air to form oxides. As we said earlier magnesium burns brightly to form magnesium oxide. Other metals which react violently with oxygen are potassium, sodium and calcium.

METALS AND DILUTE ACID

In the previous section we saw that metals react with acids to produce hydrogen gas. Again, it depends where they are in the reactivity series.

- Sodium, potassium and calcium would react very violently with acids and the reaction would be too dangerous to carry out in school. Remember, a very fast reaction is an explosion.
- Magnesium and zinc react fast with dilute acid and this is often used as a way of producing hydrogen gas for experiments.
- Iron and lead react slowly with dilute acid.
- Copper, silver and gold have no reaction with dilute acid.

DISPLACEMENT REACTIONS

A reactive metal can displace a less reactive metal from a compound. Think of it like this:

When we add an iron nail to copper sulphate solution, the more reactive iron displaces the less reactive copper from the copper sulphate compound. We end up with a new compound: iron sulphate.

We can summarise it like this:

iron + copper sulphate → copper + iron sulphate

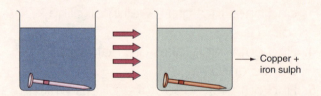

→ Copper + iron sulph

When metals are mined they are usually not found in pure form and often have to be displaced from compounds before they can be used. This is explained in the section on metals and ores (see pages 102).

ACTIVITIES

QUESTIONS

1. What is produced when sodium reacts with water?
2. Which alkali would be produced if potassium reacted with water?
3. Which is the most reactive metal in the series?
4. Which is the least reactive metal in the series?
5. Which three reactions are used to place the metals in order of reactivity?
6. Why do some metals go dull again after they have been polished?
7. Write a word-equation for zinc displacing copper from copper sulphate.
8. Why do we not react sodium with acid?

AND NOW TRY

1. There are some metals that are even more reactive than potassium. Look for the Brainiac video on alkali metals.
2. Make up a mnemonic for the reactivity series.
3. Anodising is a process used to increase the thickness of the oxide layer on aluminium. Find some pictures of anodised aluminium products.

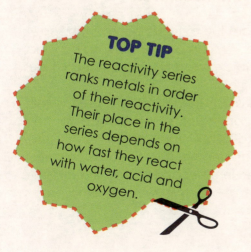

TOP TIP

The reactivity series ranks metals in order of their reactivity. Their place in the series depends on how fast they react with water, acid and oxygen.

MAKE THE LINK

Art – you learn how sculptures react with the air.

CDT – you learn how things are constructed by welding.

English – you learn to use the word 'react' in different contexts. Characters in stories react to what is happening. Some may react violently, others may not react at all.

DID YOU KNOW?

Zinc, iron and lead react slowly with oxygen. A coating of oxide forms on the surface of the metal, but this takes place over weeks or months. You can polish them and make them shiny but over time they go dull again because they react with oxygen from the air.

Gold has no reaction with oxygen. You don't need to polish gold as it does not react with the air and lose its shine.

111

OUR AMAZING WORLD

Thermite is a mixture of iron oxide and aluminium powder. The more reactive aluminium displaces the iron from its compound, as shown in the word-equation below:

aluminium + iron oxide → aluminium oxide + iron + energy

Once the reaction is started it produces large amounts of heat energy. The temperature can reach 2000°C, and the molten iron can be used to weld together steel.

SUMMARY ACTIVITIES

EARTH'S RESOURCES

EXTRACTING THE EARTH'S RESOURCES

1. This project looks at how we use and extract the earth's resources. It makes links to history, politics, music and language.

   In a cavern, in a canyon,
   Excavating for a mine
   Dwelt a miner, forty-niner,
   And his daughter Clementine

You may have heard or sung this song when you were younger. If not, look for it on You Tube. Its about a miner – obviously!

DISCUSS AND RESEARCH
- what period in history does it refer to? What was being mined?
- the men who worked in mines did hard and dangerous work, and in these conditions, music often develops. Sailors gave us traditional sea songs and the slaves in the Southern US gave us blues and jazz. What music is associated with miners?

2. When we think of mining we may think of people going down deep shafts under the earth.

Modern mining often does not involve deep shafts. It is referred to as 'surface mining' or 'strip mining'.

In groups, discuss and research:

- what is the difference between strip mining and deep mining?
- when are they used?
- compare their impact on the environment
- what is the impact of the mining of forests in Brazil and Australia?
- compare their safety for the workers
- what are placer deposits and how are they mined?

Present your findings as a poster.

3. People are now much more aware of the impact of the things they buy and use. We see statements like:

- Made from 100% recycled paper
- Fish from sustainable stocks
- Responsibly sourced

In your groups, discuss what these statements mean.

4. We should also be aware of the source of other things that we buy: where things are mined.

- In your groups, discuss and research, either:
- what are blood diamonds?
- what is dirty gold?

5. Imagine you find gold deposits in your garden, or an oil deposit in a field you own. Would these materials belong to you? Check it out: you might be surprised. The ownership of mineral rights is very complicated and involves international treaties.

DISCUSS AND RESEARCH

- how are the oil rights in the North Sea shared? Make a map showing how the right to explore is shared among various countries.
- what are some of the issues surrounding the granting of drilling licences in Alaska?
- how are the rights to ocean explorations shared in other parts of the world? Have a look at http://www.southchinasea.org/index.html.

112

ACTIVITIES

1. Match the key words with their meaning.

Keywords:
  a) element
  b) compound
  c) electrolysis
  d) solute
  e) crust
  f) plankton
  g) ore
  h) pH
  i) effervescence
  j) metamorphic

Meanings:
  1. a compound containing metal found in the earth's crust
  2. the measure of how acid or alkaline a substance is
  3. rock which has been changed by heat or pressure
  4. a substance containing only one type of atom
  5. the top layer of the earth
  6. a substance where two or more types of atoms are joined together
  7. micro-organisms which live on the ocean's surface
  8. a solid which dissolves in a liquid
  9. the separating of a compound using electricity
  10. the production of bubbles in a chemical reaction

113

2. When water is heated, it dissolves more solids. We say that the solubility of substances increases as the temperature increases. The table below shows how much potassium nitrate will dissolve in 100cm³ of water at different temperatures.

Temperature (°C)	Weight of chemical dissolved (g)
0	20
20	40
40	70
60	110
80	180

Use the data from the table above to create and complete a line graph.

Complete the following mindmap.

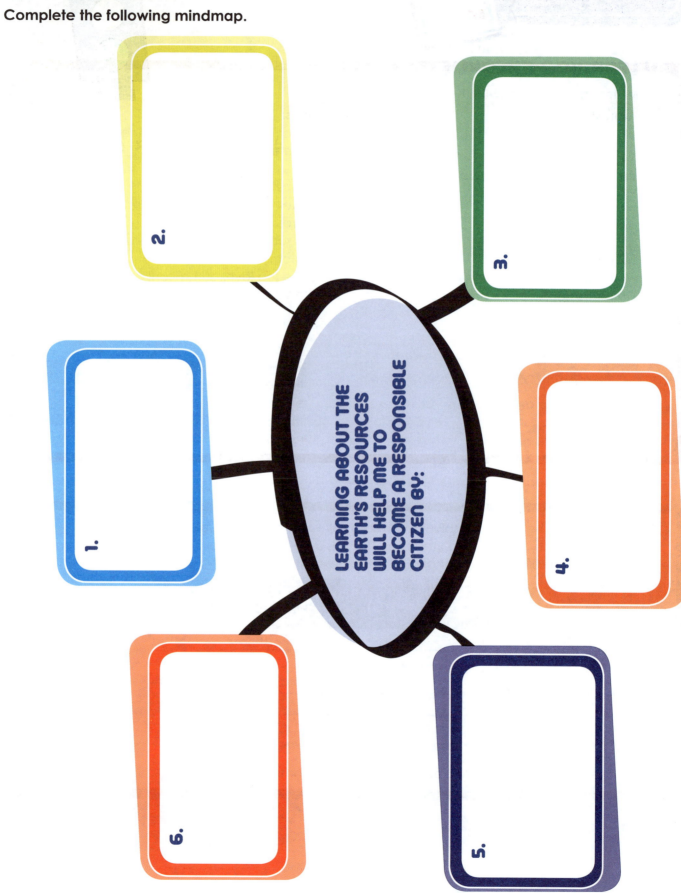

LEARNING ABOUT THE EARTH'S RESOURCES WILL HELP ME TO BECOME A RESPONSIBLE CITIZEN BY:

1.

2.

3.

4.

5.

6.

114

MY PROGRESS

MY LEARNING CHECKLIST

	Help needed	Getting there	Good to go!
1. I have studied the periodic table and can extract information from it.	⬭	⬭	⬭
2. I can use the words 'element', 'compound' and 'mixture' correctly and in context.	⬭	⬭	⬭
3. I can identify the elements in a compound and describe how their properties are different from those of the compound.	⬭	⬭	⬭
4. I can describe methods for separating mixtures and compounds.	⬭	⬭	⬭
5. I have investigated solubility and the use of non-aqueous solvents.	⬭	⬭	⬭
6. I can describe the rock types that make up the earth's crust.	⬭	⬭	⬭
7. I can explain how metals are extracted from their ores.	⬭	⬭	⬭
8. I have investigated the properties of acids and alkalis and can measure their strength on the pH scale.	⬭	⬭	⬭

115

INNOVATION AND INVENTION

SCOTTISH SCIENTISTS AND INVENTORS

In Scotland we are proud of the scientists and engineers who have made important discoveries and developments. Here are a few of them.

JAMES WATT (1736–1819)

Improved and developed the steam engine. He did not in fact invent it but developed an existing idea. The steam engine powered the industrial revolution.

CHARLES MACINTOSH (1766–1843)

Invented a waterproof material produced by sandwiching a layer of rubber between two layers of cloths. People still talk about their waterproof coats as macintoshes or macs.

JAMES YOUNG SIMPSON (1811–70)

Discovered the use of chloroform as an anesthetic. Up until then surgery had to be performed with the patient conscious.

JOHN BOYD DUNLOP (1840–1921)

Developed the modern air-filled tyre which is used on almost all road vehicles.

ALEXANDER GRAHAM BELL (1847–1922)

Invented the telephone.

SIR ALEXANDER FLEMING (1881–1955)

Discovered the antibiotic drug penicillin which is described in the section on fighting infection.

JOHN LOGIE BAIRD (1888–1946)

Invented television; or did he?

SIR ROBERT WATSON-WATT (1893–1973)

Developed radar, which is explained in the section on the electromagnetic spectrum (see pages 62).

THOMAS EDISON (1847–1931)

Probably no one has invented more things than the American inventor Edison. In his lifetime he patented 1093 inventions. The best known are the filament electric light bulb, the motion picture camera (which he called a 'kinetoscope') and the phonograph.

AND NOW TRY

1. Try typing 'inventions that changed the world' or 'ideas that changed the world' into a search engine.
2. Different cultures have different views of the past. Who did invent television? Scottish people will tell you it was John Logie Baird. You might get a different answer if you asked an American, or a Russian! Do some research and see what you think.
3. Sometimes luck and accident play a part in important scientific discoveries. Find out what led Alexander Fleming to make his world-famous discovery.

116

INNOVATION AND INVENTION

THOMAS EDISON

The phonograph invented by Thomas Edison was the first method of recording sound. Speaking into the mouthpiece in the middle caused the diaphragm and the needle to vibrate. This scratched a pattern of the sound vibrations onto the cylinder. To play the sound back, the needle moves back through the pattern on the cylinder and recreates the original sound.

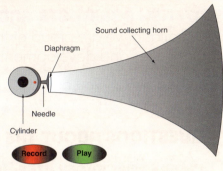

Edison's first recorded message was 'Mary had a little lamb', and if you search You Tube you will be able to listen to Thomas Edison. This method of recording and playing back sound was used everywhere until digital recording became available in the 1980s.

Like many famous men, Edison is remembered for a number of quotations. Here is one of the best known: **GENIUS IS 1% INSPIRATION, 99% PERSPIRATION**

When developing the light bulb Edison tested more than 3000 materials for a filament before he found one that satisfied him. Discuss what you think Edison's quotation means.

PATENTS AND INTELLECTUAL PROPERTY

When someone invents something new the first thing they do is patent it. Patents are the government's way of giving someone ownership of their idea. For a period of time, usually twenty years, the patent owner can control how their invention is used and to earn money from it. Patents are a form of intellectual property. We all know what property is: the things you own, like your phone, your bicycle and so on. Intellectual property could be a story you wrote, a film you made or a tune you composed. Patent and copyright laws are ways of protecting intellectual property. This is very important to business in the age of information technology, and it is intellectual property that has made Microsoft one of the biggest corporations in the world and Bill Gates one of the richest men.

THE HUMAN GENOME PROJECT

Most advances in science are not brought about by individuals but by teams of people working together. These teams are often drawn from many nations and are an important way not just of advancing science but of building understanding between different peoples and cultures. The Human Genome Project was one very big project which involved the European Union (EU) and another eighteen nations working together. The project was launched in 1990 and expected to last fifteen years. In fact, rapid advances in technology meant that it was completed by 2003. The main aims were to:

• identify more than 20 000 genes that make up human DNA
• find the pattern of the three billion pairs of proteins that make up human DNA
• store this information in a computer database
• make the information available to scientists everywhere for analysis

117

SCIENCE AND SOCIETY

There are items about science in every newspaper and often on TV and radio news. Frequently they raise questions about moral and ethical issues such as you might discuss in RME. They can also involve the government making decisions about climate change, energy and how to spend tax payers' money. On this page are a few political and ethical issues for science, which you can discuss in pairs or groups or with your teacher as a class.

QUESTIONS ABOUT THE ORIGIN OF LIFE

As we mentioned in the section on earth's oceans many scientists believe that life began in oceans. Over a billion years ago single cell organisms appeared. Fish evolved nearly 500 million years ago, followed by amphibians, reptiles, mammals and birds. The first human-like creatures evolved less than five million years ago. Scientists have collected evidence to support this theory, which was first developed by Charles Darwin in his famous book *The Origin of Species*. The dates for evolution are approximate and not everyone agrees. Many people have religious beliefs about how the earth and life were formed, which are not always compatible with this theory of evolution.

In groups, investigate and discuss:

- How do we learn about science while respecting people's religious views?
- Should religious views on the origin of life be taught alongside scientific theories?
- What are some of the faith-based views on the origin of life?
- This conflict of ideas is causing some real problems in some areas of the USA. Find out more about this.

QUESTIONS FOR THE GOVERNMENT

In the twenty-first century, science projects involve cooperation between nations on projects where the budgets may be billions of pounds. The large hadron collider (LHC) was built at a cost of over £2 billion shared between twenty EU member states. The UK contributed £100 million. This is the largest science experiment ever conducted, in which particles are accelerated around a circle of over 26 km circumference.

But in our country many children attend schools which are in very bad condition and need to be replaced. Also, lots of hospitals need to be improved and people have to wait a long time for treatment. Many old people live in poverty and young people are homeless. So you might want to ask some questions.

- What do the LHC scientists hope to discover?
- What are the potential benefits for mankind?
- Could the money be spent on more important things?

Some people would tell you that the LHC cost less than £2 per person for each member of the population of the UK. Does it help to look at it like this?

In the USA the National Aeronautical and Space Administration (NASA) has asked for a budget of nearly $29 billion for 2010. Can you think of some questions to ask the directors of NASA.

SCIENCE AND SOCIETY

ENERGY SUPPLY

Should we build more nuclear power stations?

The government must plan how to supply energy to our homes and industries over the next few decades. We saw in the energy section that we hope to reduce emission of greenhouse gases, and to do this we must reduce our use of fossil fuels. At present we use nuclear power to generate about a quarter of our power. The nuclear power stations now in use are coming to the end of their useful life and need to be replaced. Here are some options:

- Replace them with renewable sources of energy. Most scientists think this will be difficult to achieve.
- Replace them with fossil fuel power stations. We have looked at the reasons why we don't want to do this.
- Build new nuclear power stations. There are good arguments for and against nuclear power – here are some of them:

Look at the reasons for and against. Do you think they are valid? You may have to carry out more research.

For	Against
It does not produce greenhouse gases such as carbon dioxide.	It produces dangerous radioactive waste which will have to be stored carefully for thousands of years.
The technology is readily available. We do not have to develop it. France gets 70% of its energy from nuclear power.	Serious accidents have happened at nuclear power stations at Three Mile Island in the USA and Chernobyl in the former Soviet Union.
It is possible to produce very large amounts of energy from one power station.	Nuclear power stations may be the target of terrorist attacks. The technology could be used to build nuclear weapons.
We would not be dependent on foreign imports of oil and gas.	It is not renewable. Uranium fuel is a scarce resource and could run out in 50 years.

119

SCIENCE AND SOCIETY

STEM CELL RESEARCH

'Stem cells are the master cells of the human body. They can divide to produce copies of themselves and many other types of cell. They are found in various parts of the human body at every stage of development from embryo to adult. Stem cells taken from embryos that are just a few days old can turn into any of the 300 different types of cell that make up the adult body.'

By carrying out research with stem cells scientists hope to find ways to repair or replace damaged human tissue. Stem cell research may provide cures for many diseases, including diabetes, Alzheimer's disease, spinal cord injury and heart disease.

But stems cells come from human embryos. Some questions to explore and discuss:

• How are the human embryos obtained?
• What happened in the USA regarding stem cell research when George W. Bush was president?
• What happened when Barack Obama became president?

THE ROYAL SOCIETY

The Royal Society is the UK's national academy of science and is one of the most prestigious science organisations in the world. To become a Fellow of the Royal Society is one of the highest honours a scientist can receive. Have a look at their website to see what they think are the important science issues of today: http://www.royalsoc.ac.uk/Science-Issues/.

SUMMARY ACTIVITIES

SCIENCE NEWS

How science is reported in the news is vitally important. Responsible citizens need to have accurate information to make decisions and be effective contributors to society. All news media employs science correspondents to work in this area. In this project we will use a wide range of skills to produce a science newspaper. So start by thinking of a name for your paper and design a headline. Look at some daily papers for ideas. Here are some of the things you might include – you will need a big team to share all the tasks:

BREAKING NEWS
Most papers have a front page article with a new story. You can find something in a daily paper, on TV or on the internet. Try a search for 'breaking science news' or look at http://www.newscientist.com/section/science-news for some ideas. Remember, you will need some pictures.

AN OPINION PIECE
There are a lot of areas in science where people disagree, so let's have a short article where someone argues their views on something like: nuclear power, wind turbines, stem cell research. Anything you like!

TECHNOLOGY REVIEW
Someone in your group might have a new phone, a new MP3 player or some other piece of technology. Ask them to write about it: describe it, say what they like and don't like about it and maybe make suggestions for improving it.

AN INTERVIEW
Conduct an interview with someone who works in a branch of science. Ask them some questions about their work and write a profile.

WEB STUFF
Compile a list of websites that give interesting and fun science activities for children and young people.

TV REVIEW
Watch a TV programme about science and write about it.

THE NIGHT SKY
Lots of papers have a column which tells people what to look for in the sky at night. You could use www.**stellarium**.org/ to give you some ideas.

121

ACTIVITIES

1. Working in pairs, choose a Scottish scientist to find out more about. It could be one of the scientists listed on page 116, or another one whom you have heard about or come across in your research.
   Find out about the scientist's life, his/her research and the contribution that it makes to the world. Present your findings in a Powerpoint presentation to the rest of the class.
2. Pick a science topic that raises ethical questions. Research and discuss both sides of the debate, then write a short essay which summarises all the issues and presents your own opinions.
3. Organise a class debate on stem cell research, using the information on page 120. Half the class should present the case for stem cell research and the other half should present the objections. Both sides can ask questions and challenge the other side!
4. Research careers in science and engineering. What career paths are available? What training would you need to be a scientist or an engineer? Are there any jobs in science that would appeal to you?

Complete the following mindmap.

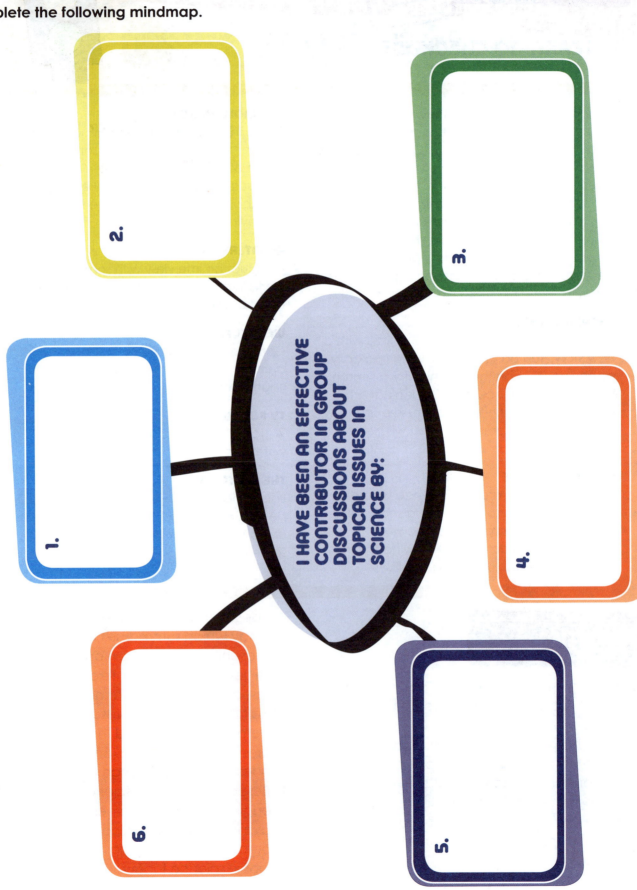

I HAVE BEEN AN EFFECTIVE CONTRIBUTOR IN GROUP DISCUSSIONS ABOUT TOPICAL ISSUES IN SCIENCE BY:

1.

2.

3.

4.

5.

6.

MY LEARNING CHECKLIST

	Help needed	Getting there	Good to go!
1. I have worked with others to find and present information on how Scottish scientists have contributed to research and development.	◯	◯	◯
2. I have evaluated how science is presented in the media.	◯	◯	◯
3. I have discussed ethical issues around scientific research.	◯	◯	◯

ANSWERS

PLANET EARTH

ANIMALS AND THEIR HABITATS

1. Long eye lashes, hair to protect ears and nostrils that can close.
2.

Adaptation	Purpose
Layer of blubber	Insulation
Black skin	Absorb heat from sun
Pads on soles of feet.	Grip on ice
Curved claws	Catch and hold prey, dig in snow
Large feet	Spread weight on snow, swim fast.

3. Good eyesight, sharp talons to kill prey, sharp beak for tearing flesh.
4. The prey needs to keep an all round lookout for predators.

PHOTOSYNTHESIS

1. Light energy to chemical energy
2. Water and carbon dioxide
3. Glucose and oxygen
4. They make it by photosynthesis
5. By eating plants and other animals
6. Respiration

CHEMICALS AND AGRICULTURE

1. Nitrogen, phosphorus, potassium.
2. Plants need compounds of nitrogen.
3. By bacteria in plant roots or by lightning.
4. Animal waste and remains are broken down by bacteria.
5. The production of greenhouse gases.
6. They tend to lead to the destruction of soil structure and wash off into the rivers and oceans, where they reduce dissolved oxygen.

ENERGY TRANSFER AND BUILDINGS

1. Conduction, convection and radiation.
2. Dark surfaces.
3. Light surfaces absorb less radiation.
4. Thermals.
5. Loft insulation, double glazing, draught excluders, wall insulation.
6. Radiation.

RENEWABLE ENERGY

1. The sun.
2. Potential → kinetic → electrical.
3. Light → electrical.
4. Passive solar heating.
5. The moon and the sun.
6. Damage to wildlife and spoiling the landscape.

FOSSIL FUELS

1. Coal, oil and natural gas.
2. Coal.
3. New supplies are being discovered.
4. It will rise.
5. Over 100 million years ago.
6. They cannot be replaced.
7. A rock cap through which they cannot pass.

CHANGES OF STATE

1. a) Melting.
   b) Solidification.
   c) Evaporation.
   d) Condensation.
   e) Sublimation.
   f) Deposition.
2. More energy is required to raise the temperature of water.
3. (Hoar)frost
4. Because Scotland is surrounded by water.

HEATING AND COOLING MATTER

1. Solid, liquid and gas.
2. It makes them move faster.
3. It expands.
4. It contracts.
5. c) About the same (the volume does in fact increase slightly).
6. c) About the same (see 'Did you know?').
7. The spacing increases because the steam takes up much more space than the water.

AIR PRESSURE

1. Pressure is equal inside and outside our bodies.
2. Pressure outside your body is now greater than inside.
3. Vacuum.
4. Pressure will increase as the particles move faster.
5. Diving or swimming underwater.

GASES OF THE AIR

1. Oxygen.
2. Carbon dioxide.
3. Glowing splint test.
4. Lime water test.
5. It is more efficient and prolongs their life.
6. Unreactive elements do not take part in many chemical reactions.
7. Oxygen is reactive because it takes part in lots of chemical reactions.

CLIMATE CHANGE

1. Carbon dioxide.
2. Photosynthesis and respiration.
3. Burning of (fossil) fuels.
4. No, it has always happened but human activity is upsetting the balance.
5. Carbon dioxide will increase as there are fewer plants to use it in photosynthesis.

THE SOLAR SYSTEM

1. Terrestrial and gas.
2. Orbits.
3. Pluto.
4. The moon.
5. They make things look too close together.
6. Nuclear reactions.

BEYOND THE SOLAR SYSTEM

1. Planet, solar system, galaxy, universe.
2. Eight minutes.
3. Four years.
4. Andromeda.
5. 100, 000 light years.
6. The distance light travels in one year.

FORCES, ELECTRICITY AND WAVES

FRICTION

1. The opposite direction to the motion.
2. Kinetic energy to heat energy.
3. Roughness in surfaces catching each other.
4. By lubrication.
5. In a bicycle chain, car engine etc.
6. Picking things up, between shoes and the floor etc.
7. The ice reduces the friction between the tyres and the road.

FRICTION AND CARS

1. Frictional force due to air.
2. Brakes and tyres.
3. By snow, ice or rain.
4. It increases.
5. To reduce fuel consumption.

FIELD FORCES

1. Field forces and contact forces.
2. Magnetic, electrostatic and gravitational.
3. Cling film, photocopiers and inkjet printers.
4. Electric motors.
5. It is very much less.
6. 50 N.
7. It would be less.
8. Weightlessness.

DENSITY AND FLOATING

1. Density = mass/volume.
2. g/cm^3.
3. $1 g/cm^3$.
4. No, its density is greater than water.
5. $0.8 g/cm^3$.
6. a) $240 cm^3$
   b) $0.75 g/cm^3$

BATTERIES

1. Chemical to electrical.
2. Electrical to chemical.
3. A non rechargeable battery.
4. A rechargeable battery.
5. Two electrodes and an electrolyte.
6. A battery is two or more cells joined together.
7. 4.

CIRCUITS

1. Negative
2. Negative to positive.
3. Provides energy.
4. Electrical to heat and light
5.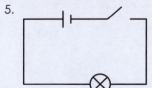

ANSWERS

6.
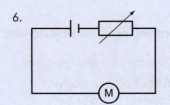
7. Electrical to kinetic.

CIRCUIT LAWS

1.
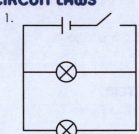

2.
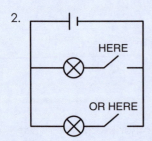

3. 0·4 A
4. 0·3 A
5. 3 V
6. 6 V

ELECTRONIC SYSTEMS

1. Input, process, output.
2. The screen.
3. The keypad.
4. Thermistor and thermocouple.
5. Light-dependent resistor or solar cell.
6. Electrical.
7. Screen, printer, speakers etc.
8. Keyboard, mouse, disc drive, wi-fi, network etc.

BODY SYSTEMS

BREAK IT DOWN

1. Oxygen, heat and dissolved chemicals.
2. To pump blood round the body.
3. It collects oxygen.
4. Carbon dioxide.

SOUND

1. Compressions.
2. 20Hz, 20 kHz.
3. Ultrasound.
4. Dolphin and bat.
5. Elephant.
6. 340 metres per second (m/s).
7. 1020 metres.

REFLECTION AND REFRACTION

1. Normal.
2. They are equal.
3. Speed reduces and direction of travel changes.
4. Blue light.
5. Spectrum.
6. Convex and concave.
7. It acts as a magnifying glass.
8. A camera.

ELECTROMAGNETIC SPECTRUM

1. 300 000 000 m/s.
2. Gamma rays.
3. Radiant heat.
4. Ultrasound.
5. Causes sunburn/skin cancer.
6. Wilhelm Röntgen.
7. Radioactive materials.
8. One millionth of a millimetre.
9. Radar.
10. To kill cancer cells, produce images of internal organs.

EYES AND EARS

1. On the retina.
2. Iris.
3. Focuses light to form an image.
4. Along the optic nerve.
5. Light to electrical.
6. Hammer, anvil and stirrup.
7. In the cochlea.
8. To collect sound waves.

5. Rib cage and diaphragm.
6. Through the oesophagus.
7. Nutrients.
8. Water.
9. Through the anus.

TECHNOLOGY AND HEALTH

1. a) To kill cancer cells.
    b) To take pictures of the inside of the body (especially bones).
    c) To treat muscle injuries.
    d) To treat skin conditions like acne.
2. It builds up a 3D image.
3. A long thin piece of glass.
4. To listen to sounds inside the body.
5. 37°C
6. There is less blood loss.
7. An endoscope.

CELLS

1. Nucleus, cell membrane, cytoplasm.
2. Cell wall, vacuole, chloroplasts.
3. a) Stores information and controls the cell.
    b) Allows substance to pass in and out of the cell.
    c) Chemical reactions take place here.
4. Vacuole and cell wall.

MICRO-ORGANISMS

1. Fungi, bacteria.
2. Plague, leprosy, tuberculosis.
3. In the intestine.
4. Cheese, yogurt.
5. By keeping it cool, in a fridge.
6. Mushrooms, yeast.
7. Athlete's foot, dry rot, mould on food.
8. By breaking down animal waste and remains of dead plants and animals.

MATERIALS

ELEMENTS

1. A substance containing only one type of atom.
2. Bromine and mercury.
3. Periodic table.
4. Halogens.
5. Noble gases.
6. Alkali metals.
7. Metals from non-metals.

MIXTURES AND COMPOUNDS

1. There is an energy change and something new is formed.
2. Molecule
3. It only contains the two elements named
4. It contains the two elements names and oxygen.

DEFENCE

1. Through a cut, by breathing or by swallowing with food and drink.
2. 37°C.
3. Flush something harmful out of your body.
4. By surrounding and digesting them or by producing antibodies.
5. To make it hard for the bacteria or virus to survive.

FERTILISATION

1. Zygote.
2. Embryo.
3. Embeds in the lining.
4. Amniotic fluid.
5. Food and oxygen.
6. Waste products.
7. In the placenta.
8. Nicotine, drugs, alcohol.

GENETICS

1. In the nucleus of cells.
2. Double helix.
3. Twenty-three.
4. A, T, C and G
5. Blood, hair, semen, skin.
6. Twenty-three from the mother (egg) and twenty-three from the father (sperm).
7. Genetic fingerprinting.
8. Sickle cell anaemia, cystic fibrosis.

5.

Elements present	Name of compound
Magnesium and chlorine	Magnesium chloride
Sodium and fluorine	Sodium fluoride
Copper, nitrogen and oxygen	Copper nitrate
Potassium, chlorine and oxygen	Potassium chlorate

ANSWERS

SEPARATING COMPOUNDS

1. Heat
2. Energy
3. Ultraviolet
4. Silver
5. a) Copper and chlorine
   b) Green crystalline solid
   c) Chlorine, a greenish poisonous gas; copper is a brown metal
6. Electricity
7. Compounds of silver and halogens

DISSOLVING

1. Solute – salt; solvent – water; solution – salty water.
2. Increase temperature; decrease particle size; stir.
3. The particle size; volume of water; weight of solid; level of stirring (any two are OK).
4. Small crystals.
5. Saturated solution.

THE EARTH

1. Molten iron and nickel.
2. In the mantle.
3. Magma bubbles through the earth's crust.
4. Oxygen and silicon.
5. Movement of the tectonic plates.
6. Igneous.
7. Metamorphic.
8. Sedimentary.

EARTH'S OCEANS

1. Photosynthesis by green plants and phytoplankton.
2. Phytoplankton.
3. Zooplankton.
4. 70%.
5. Magnesium.
6. Placer deposits.
7. From waves and tides.
8. Gold, titanium, tin, diamonds.

METALS AND ORES

1. A compound of a metal with other elements.
2. Gold and silver.
3. Iron.
4. Iron ore and carbon.
5. copper + iron oxide → iron + carbon dioxide.
6. Ilmenite.
7. Electrolysis.
8. It is too reactive.

CHEMICAL AND PHYSICAL CHANGES

1. Bubbles (effervescence); colour change; solid forming and sinking to the bottom (precipitation).
2. Increase temperature, decrease particle size, increase concentration.
3. Rusting; digesting food; baking a cake; chalk bubbling in acid; magnesium burning.
4. Speed up chemical reactions.
5. Enzyme.
6. The atoms combine – join together.
7.

Physical	Chemical
water evaporating	gas burning
ice melting	iron rusting
sugar dissolving	egg frying

ACIDS AND ALKALIS

1. Weak acid.
2. Neutralisation.
3. 1 or 2.
4. Universal indicator.
5. Red.
6. Strong alkali.
7. Green.
8. Concentration.

MORE ABOUT ACIDS

1. Hydrogen
2. Zinc + hydrochloric acid → zinc chloride + hydrogen.
3. Carbon dioxide.
4. Calcium, carbon and oxygen.
5. Magnesium and chlorine.
6. Limestone reacts with acid, quartz doesn't.
7. Add lime.
8. Phosphorus.

REACTIVITY

1. Hydrogen.
2. Potassium hydroxide.
3. Potassium.
4. Gold.
5. With water, acid and oxygen.
6. They form a layer of oxide.
7. Zinc + copper sulphate → copper + zinc sulphate.
8. It would be too dangerous.